H. P. Robinson, Captain Abn

The art and practice of silver printing

H. P. Robinson, Captain Abney

The art and practice of silver printing

ISBN/EAN: 9783743330030

Manufactured in Europe, USA, Canada, Australia, Japa

Cover: Foto ©berggeist007 / pixelio.de

Manufactured and distributed by brebook publishing software
(www.brebook.com)

H. P. Robinson, Captain Abney

The art and practice of silver printing

THE

ART AND PRACTICE

OF

SILVER PRINTING.

BY

H. P. ROBINSON & CAPT. ABNEY, R.E., F.R.S.

THE AMERICAN EDITION,

NEW YORK:

E. & H. T. ANTHONY & CO., NO. 591 BROADWAY.

1881.

PREFACE.

SILVER printing has been often doomed, but it still sur-
vives. Other processes of photographic printing have
been introduced, nearly all of them having their individual
merits, especially that of permanency, but all lacking in
two essential qualities—ease of production and beauty
of result. In these particulars no process has ever
approached the one to the working of which this little
book is devoted. The one defect of silver printing is the
possibility of its results fading; but surely it is better
to be beautiful, if fading, than permanent and ugly. It
is better to be charmed with a beautiful thing for a few
years, than be bored by an ugly one for ever. But is
silver printing necessarily a fading process? We have
in our possession a large number of silver photographs
produced from twenty to twenty-five years ago, which are
as perfect in tone and colour as when they were produced.
Carefully prepared, and properly kept, a silver print
should be as permanent as any other. That silver prints
should be permanent as well as beautiful, has been the
object of

THE AUTHORS.

TABLE OF CONTENTS.

CHAPTER   PAGE

I.—Preliminary Experiments 1

II.—Preparation of Albumenized Paper 6

III.—The Sensitizing Bath 13

IV.—How to Keep the Sensitizing Bath in Order ... 20

V.—Silvering the Paper 26

VI — Washed Sensitive Paper 31

VII—Cutting Paper 36

VIII.—Printing-Frames 42

IX.—Preparing the Landscape Negative 45

X.— Printing the Landscape 49

XI.—Preparing the Portrait Negative 57

XII.—Vignetting 60

XIII.—Printing the Portrait 69

XIV.—Combination Printing 74

XV.—Toning 85

XVI.—Fixing the Print 92

XVII.—Washing the Print 95

XVIII.—Printing on Plain Paper 99

XIX.— Printing on Resinized Paper 100

XX.—Printing on Gelatino-Chloride Emulsion Paper ... 103

XXI.—Drying the Prints 105

XXII.—Mounting Photographs 110

XXIII.—Defects in Prints 115

XXIV.—Encaustic Paste... 117

XXV.—Enamelling Prints 119

XXVI.—Cameo Prints 121

Appendix 123

CHAPTER I.

THEORY OF SILVER PRINTING.

PERHAPS it may be wise, first of all, to give the reader some account of the manner in which the subject of silver printing is to be treated, before entering into very minute details, so that it may be followed as a whole, instead of being studied in fragments, a course which is sure to lead to failure, from a want of comprehending what may have been skipped. To understand "the why" and "the wherefore" of every detail is an essential in most occupations, and it is wonderful that photographers are satisfied with the results of rule-of-thumb formulæ, instead of reasoning out their utility. In the following pages most of the theoretical considerations will be brought out in such a manner that everyone will be able to understand them, provided only that there is a slight acquaintance with the name and properties of the chemicals which are dealt with.

PRELIMINARY EXPERIMENTS.

Into a glass beaker put a couple of pinches of common salt, which must be dissolved in a little water.

In a test-tube* dissolve about an equal amount of silver nitrate

---

* Such things as test-tubes should be found in every photographer's work room ; they cost little, and are always useful for working solutions. The sizes recommended are $\frac{3}{4}$-inch, $\frac{1}{2}$-inch, and 1-inch diameter. A dozen of each will not be out of the way.

(Ag NO₃), and add it to the salt solution. We shall find that we have an immediate precipitate, for chloride of silver will be formed by what is called double* decomposition, and there will remain in solution a soluble salt known as sodium nitrate. When the silver chloride has settled down, decant off the liquid, and add water to it once or twice, draining off each time. Divide the chloride into four parts, placing each part on a strip of glass. On two of them pour a little common salt solution, and on the other two pour a little solution of silver nitrate; take one of each pair, and place it in a dark cupboard (if warmed, the quicker will be the operation) to dry. Take the other two moist portions of chloride into the open air, and expose them to day-light, and note the results. It will be seen that one of these will darken very rapidly to a violet colour, whilst the other will re-main much lighter, though perceptibly blackening. After a time the latter will appear to grow deeper, whilst the former will become a deep black. The one that blackens most rapidly will be found to be that one on which the silver nitrate was poured. Divide the slightly blackened chloride on the strip of glass into two portions, and over one pour a little beer, and over the other a weak solution of potassium nitrite, and again note the difference. It will be found that here the blackening commences anew, but proceeds much more rapidly on that portion over which the nitrite was poured. Here are the experiments. What do they teach?

Potassium nitrite, and silver nitrate, are both inorganic salts, and they both have an affinity for—that is, tend to combine with—any of the halogens (by which are meant such bodies as chlorine, iodine, bromine, and fluorine). In the former case we have silver chloride formed with a little hypo-chlorous acid; in the latter we have a more difficult decomposition : the potassium

Sodium Chloride	and	Silver Nitrate	form	Silver Chloride	and	Sodium Nitrate
* Na Cl	+	Ag NO₃	=	Ag Cl	+	Na NO₃

nitrite is decomposed into hydrochloric acid and potassium nitrite.*

We can tell that chlorine is liberated by the action of light on silver chloride, since if we prepare some as above, well wash it, and expose it to light in pure water, we shall find that the latter contains chlorine, since a few drops of silver nitrate poured into it after exposure give a white precipitate.

If we make the same experiments with the dried portions of silver chloride as we did with the moist, we shall obtain the same results, with the exception that with the dried, in which there is excess of salt, there will be hardly any discolouration. The experimentalist should also note that if the darkened chloride be broken up, the interior retains its white colour in all its purity. This tells us that the discolouration is *almost* confined to the surface, hence it is useless, for printing purposes, to have such a mass of chloride as would be opaque, since all but a very thin film would be unacted upon.

If the darkened chloride be examined closely, it will be seen that the colour varies, being bluer in the case of that which has silver nitrate in contact with it (either moist or dry) as compared with that which is darkened in contact with the potassium nitrite. We have the best of reasons for believing that the blue colour is really due to a combination between the sub-chloride and the oxygen contained in the water or in the air. The true colour of the sub-chloride is that which is exposed beneath an oxygen absorbent such as the nitrite.

Practical printers are aware that albumenized paper containing a chloride is employed for producing silver prints, and the pro-

---

Chlorine and Potassium Nitrite and Water give Hydrochloric Acid and Nitric Acid

* $2Cl + K NO_2 + H_2O = 2HCl + K NO_3$

and

Chlorine, Silver Nitrate, and Water give Silver Chloride, Hydrochlorous Acid, and Nitric Acid

$2Cl + Ag NO_3 + H_2O = Ag Cl + HClO + HNO_3$

bability is that the albumen must exercise some kind of influ-
ence on the resulting picture. Let us examine this, and see what
effect it can have. Carefully break an egg, and separate the yolk
from the white, pouring the latter into a beaker. Beat up the
white with a bundle of quill pens, allow the froth to subside, and
then filter it. Pour a little of the filtered albumen (the white of
egg) into a test-tube, and add a little silver nitrate solution
to it, and expose the precipitate which falls to light. It will be
seen that it darkens rapidly, assuming a foxy red colour. Take
a couple of glass plates and coat them with plain collodion, wash
under the tap, and whilst still moist flow albumen over them
two or three times, and set them up to dry. When thoroughly
dry, plunge them for a few seconds into a weak solution of silver
nitrate (30 grains to the ounce of water will suffice), wash one
under the tap, and then allow both to dry again. Take both
plates out into the light, and note the results. The one from which
the silver nitrate has not been washed will darken very rapidly,
the other will take some time to start; but if the exposure be
sufficiently prolonged, it will gradually assume a hue equally as
dark as the other.

If we repeat these experiments with gelatine, which is used
as a sizing in some papers, we shall find very much the same
nature of things taking place, the differences being so slight,
however, as not to require detailed notice.

So far, then, we have considered the darkening properties of
the silver compounds which are to be used by the printer, but
it remains to be seen what *permanency of darkening* they possess.
If we treat the darkened silver chloride solution exposed with
the silver nitrate or the potassium nitrite to a solution of hyposulphite of soda or ammonia, both of which are solvents of the
white chloride, we shall find that a residue of metallic silver is
left behind. If we treat the darkened albuminate of silver with
the same agents, we shall find that very little change is effected
by them.

From this we may gather that the action of light on them is of a totally different nature.* This is also most marked if we treat the two with hydrosulphuric acid solution (sulphuretted hydrogent). It will be found that the colour of the darkened silver chloride becomes more intense, while the other is bleached, or, rather, becomes of a yellow tint. This last effect has an important bearing on the permanency of silver prints, as will be more fully explained when considering the subject of fixing the print.

---

* With the former we have this action—

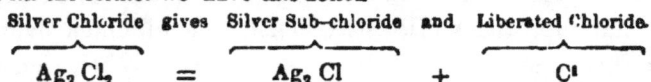

Silver Chloride gives Silver Sub-chloride and Liberated Chloride.

$$Ag_2Cl_2 = Ag_2Cl + Cl$$

With the latter the silver in combination with the organic matter, which is in a state of oxide, is probably reduced to the state of sub-oxide.

† Sulphuretted hydrogen may be prepared by pouring dilute sulphuric acid on ferric sulphide. The chloride or the silver compound, when damped, may be held over it, taking care that no liquid is spirted up on to it.

CHAPTER II.

PREPARATION OF ALBUMENIZED PAPER.

In printing on albumenized paper we must divide the operations, and give a detailed account of each. In case the reader may desire to prepare his own paper, we give the following formula and directions.

To prepare the albumen, procure a sufficient number of eggs, remembering that the white of a large egg will be about a fluid ounce ; have a cup to collect the yolks, and a four-ounce measure at hand. Give the centre of the egg a smart blow against the top of the cup. The shell can now be readily pulled in two, the yolk remaining unbroken with part of the albumen in one half, and the rest of the albumen in the other half of the shell. Take the halves, one in each hand, and pour the albumen from one to the other, holding them over the small measure. As the operation continues, the yolk will gradually separate, the white falling into the vessel below. If conducted with care, the whole of the latter will be collected without breaking the yolk. If the yolk break, some will be sure to find its way into the measure along with the white, and this, together with the white speck known as the tread, must be rigorously taken out by means of a spoon. The *uncontaminated* white is then poured into a large jar. If the operator carefully collect the white of each egg into the four-ounce measure first, he will find his labour much diminished, as it is awkward to get out the small pieces of yolk from a large quantity of albumen. The eggs are thus broken, and the white collected till there is a sufficient quantity for the purpose in hand. Suppose we are going to make up 20 ounces of solution, then about 18 ounces of white of egg must be found in the jar. One point to settle is the amount of salt to be used to each ounce of albumen. It must be recollected that a medium quantity is the best for medium negatives; anything between 20 and 40 grains per ounce may be used. We prefer ourselves about 25. Supposing this quantity to be used, we proceed to dissolve 500 grains of chloride of ammonium in 2 ounces of water, and add it to the albumen. It has been proved that as regards colour of the picture, it does not matter what chloride is used. To prevent crystallization, it is better to use ammonium, which contains a great r amount of chlorine than do sodium or potassium chlorides. It must now be beaten up till it is in a froth. This breaks up the fibrous matter, and on subsidence the liquid will be found to be limpid. The most convenient implement with which to beat up the albumen is the American egg-beater. Three or four minutes' work is quite sufficient to make the whole into a froth. An ordinary culinary whisk, such as is used in the kitchen, may also be put into requisition, or, in default of that, a bundle of quill pens. A lesson in producing a froth can be learnt from the cook of the establishment. When the salted albumen has settled it must be filtered, which, perhaps, is best effected through a sponge, though glass-wool is a capital substitute. In either case a small, loosely-fitting plug is placed in the neck of an ordinary funnel, and, after rinsing with cold water, the albumen is poured in, and allowed to filter through slowly. It is advisable to avoid bubbles as far as possible, and the accompanying arrangement will be found to avoid their formation. The funnel is placed in the position shown (fig. 1); the capillary attraction between it and the glass will cause the drops to trickle

down the side, and collect, without bubbles, at the bottom. This
little contrivance will be found of use in other operations besides

Fig. 1.

that of silver printing, and should be made a note of. The albumen may also be filtered through one, two, or three thicknesses of
muslin, according to its fineness, tied over the mouth of a bottle or

Fig. 2.

beaker of which the bottom has been removed. The albumen is
placed in a vessel slightly larger than the filter, which is allowed
to sink gradually. When full it is withdrawn, and the fluid
poured into the dish. By this plan upward filtration is estab-
lished. The fluid may be poured into the filter itself, and used
in the ordinary manner.*

On a larger scale, white of eggs in a fresh condition can be
obtained from egg merchants who utilize the yolks by selling
them to the grocers and confectioners. Albumen can be

---

* Those who prepare collodio-albumen plates will find the upward filtration arrangement of immense value, as bubbles are unknown by it.

obtained by the gallon in this condition, according to the price of eggs. It will be evident that there is considerable economy in taking the whites wholesale. As a rule, about three gallons of albumen will coat two reams of albumenized paper. Mr. England (to whom we are indebted for so many of our remarks on albumenizing paper) procures about the latter quantity at a time, and beats it up mechanically in a large vat holding some fifty gallons, in order to allow space for the froth. He allows the albumen to rest four days before employing it, and filters it through three thicknesses of flannel.

The quality of paper to be used varies considerably with the custom of the printer. Thus, in some countries, we find a much thinner paper used than in England. The great desideratum is that it should be perfectly opaque to transmitted light. A good test of this is to make a couple of black ink marks on a piece of white paper, and then press down firmly the paper it is proposed to employ over this. If the black ink marks are indistinguishable, the paper will do as regards this quality, as the light reflected from the surface which gives the impression of whiteness to the eye is much stronger than the light which penetrates through it, and is absorbed by the black lines. As to quality, it is best to trust to the manufacturer, those known as Saxe and Rives papers answering better than any other that we know of. The Rives is, when moist, a paper which is more easily torn than the Saxe, and, consequently, we recommend that the former be employed for small work, such as portraits, and the latter for large landscape prints.

In regard to the sizes to be albumenized, it must be left to the operator to say what will be the most useful to him. It is rarely advisable to albumenize less than a half sheet of paper, the whole size of which is about 22 by 18 inches; 11 by 18 is not an inconvenient size to manipulate. At any rate, a dish larger each way by a couple of inches than the paper must be procured, and put on a level table. The temperature of the room should be at, least 90°, in fact, the hotter it is the more glossy will be the resulting paper. The solution, free from bubbles, is poured in, and should be of a depth of at least $\frac{1}{2}$ an inch. Suppose the smaller size to be coated, before commencing, the paper is taken by the two opposite corners, the hands brought together, and

Fig. 3.

the convex side brought on to the surface of the fluid; the hands are then separated, and the paper will gradually float on the surface. One corner should be gradually raised to see that all air-bubbles are absent. If there be any, they should be broken with the point of a glass rod, and the paper again lowered. Bubbles can usually be seen through the paper, and, instead of raising it, a few gentle taps with the finger over the spot will generally move the bubble to the edge of the paper. In practice, some have found it well to moisten the surface of the paper with a damp sponge, and when quite surface dry to albumenize it. This should, however, be unnecessary. The sheet should remain on the albumen a little over a minute, when it could be gently raised by one corner and allowed to drain over a basin; it is then caught by a couple of American clips and hung up to dry.*

Supposing a whole sheet is to be coated, it will be found more convenient to take the sheet by the corners of *one end*, one in each hand, and to lower the surface near the end of the dish, and gradually draw the paper over the side of the dish till the whole surface is flat. Bubbles can be got rid of as shown above.

* If bubbles are seen, they must be broken, and the sheet floated again for another minute.

Two large dishes are usually employed, and by the time the second sheet is floated in the second dish, the first sheet of paper is ready for removal from the first dish. The sheets, when slowly removed from the bath, are allowed to drain a few seconds, and then thrown over wooden rods of some two inches in diameter, which are removed to a rack, and placed near a trough to collect the drainings.* When drained sufficiently the rods are removed to other racks, and the paper allowed to dry spontaneously.

It is the practice of some albumenized paper manufacturers to hang the sheets over a line, uncoated side next the line; but this is a mistake, as it will nearly always be found, on sensitising the paper and exposing it, that a mark is left across the paper corresponding to the part where the string touched the back of the paper.

In practice we have found that each sheet of paper takes up about $\frac{1}{3}$ oz. of solution, and, of course, its equivalent quantity of salt. The principal difficulty in albumenizing paper is the occurrence of lines on the paper in the direction in which it was placed on the surface of the albumen. Any arrest of motion in floating the paper will cause them, but more usually it is due to imperfect beating up of the solution. Some papers are not readily coated with albumen, in which case the remedy given above may prove effectual; or a little solution of oxgall may be equally well applied. A want of gloss in the dried albumen may be due to too long a floating on the fluid, or to floating and drying the paper in too low a temperature. The explanation of the first cause is that albumen, when fresh, has an alkaline reaction, due to the presence of a small quantity of soda, which may be said to be its base, and any alkali will dissolve the gelatinous sizing of a paper. When the sizing is dissolved, instead of re-

---

* The drainings are added to the next batch of albumen which is prepared.

maining on the surface, the albumen sinks into the paper, and thereby the gloss is lost.

When albumen is stale it no longer possesses this alkaline reaction, but has an acid reaction quite visible on the application of blue litmus paper to it; the blue colour disappears and is replaced by a red tint. When in the alkaline state, the paper is much more difficult to coat, but an acid condition means the production of inferior tones.

Rolling the Paper.—The paper, when dried, is often rolled with a heavy pressure to improve the gloss; a copper-plate press is found to answer admirably, placing the albumenized side next the bed. This rolling should not be necessary if attention be paid to the temperature of the preparation room. The higher the temperature the finer will be the gloss, as we have already said.

CHAPTER III.

THE SENSITIZING BATH FOR ALBUMENIZED PAPER.

To render albumenized paper sensitive to light it has to be treated with a solution of silver nitrate, and the most convenient method of applying it is to float it on a dish containing the silver salt in solution. The first point to consider is the strength of the solution. If we float albumenized paper (face downwards) on a solution of 10 grains of silver nitrate to the ounce of water, we shall find, what at first sight may seem to be remarkable, that the albumen will be dissolved away from the paper, and that there will be a precipitate left in the silver solution. Why is this?

It must be remembered that albumen is soluble in water: it is coagulated or insoluble in water when combined with silver nitrate. The fact is that the quantity of silver nitrate in the solution we have been experimenting with is too small. The water dissolves the albumen first, and then the silver has time to act upon it to form the insoluble albuminate. If we soak paper in common salt, and treat it in the same way with the same strength of solution, we shall find that this is not the case: the silver chloride will remain on the paper. From this we learn two facts.

1st. That the silver solution has a greater affinity for the

chloride than for the albuminate, and that in an equal mixture
of the two more chloride would be formed than albuminate;
in other words, that the ammonium chloride would be totally converted into silver chloride long before the silver albuminate was
formed.

2nd. That a certain strength of silver nitrate is necessary
to prevent the albumen dissolving from off the paper.

This last fact has fixed the lowest strength of any sensitizing
solution to be thirty grains to the ounce, and even if this be
taken as a limit, it is necessary that the water should be
rendered less active by holding some other soluble matter in
its embraces. This is usually effected by adding some other
neutral and inactive nitrates. There does not seem to be any
theoretical limit to amount of silver nitrate in solution, but
practically it rarely contains more than 80 grains to the ounce,
though occasionally we have heard of it being used of a strength
of 100 grains to the ounce.

The important point now presents itself. How are we to fix
the strength of the bath? What principles must we follow?

To answer these questions we extract a passage from another
work of this series.*

"If a paper be coated with albumen (say) in which has been
dissolved a certain quantity of a soluble chloride, and floated on a
silver solution, both chloride and albuminate of silver are formed.
It depends, however, on the strength of the solution as to what
proportions of each are present, owing to the fact that the organic
compound is much slower in formation than the chloride, and has
less affinity for the silver. If the silver solution be not sufficiently
strong, the chloride may rob that portion of it with which it is in
contact of all the silver before any (or, at all events, sufficient)
albuminate has been formed, the molecule being composed almost
entirely of silver chloride. The stronger the silver solution the

* "Instruction in Photography," 4th edition, page 121.

more 'organate' will it contain ; whilt if it be very weak, very little will be present.   Hence it is with albumenized paper which is weakly salted with a silver chloride a weak sensitizing bath may be used, whilst if it be rich in the chloride it must be of proportionate strength."

It will now be seen that the proportion of chloride to albumen has to settle the point.   We next have to consider the time during which the silver should be in contact with the paper when the floating is commenced.   Let us take the case of a strong silver solution, and consider the action that will follow. Immediately the paper is placed in contact with the solution, silver chloride is formed, and the amount of the silver nitrate in the layer of fluid in immediate contact with the surface being scarcely diminished by the formation of silver chloride, the albuminate is formed almost simultaneously, forming a ulm which is to a great extent impermeable to the liquid.   But even before this layer is coagulated, the next layer of chloride will have been formed, so that we may say we have one layer of albuminate and chloride of silver, and one layer of chloride of silver alone.

The further penetration of the silver solution will be very slow; hence, for fully saturating both the albumen and the salt with silver, the time of flotation must be prolonged.   For some purposes, however, this is not necessary, as will be seen presently.

Next let us trace the action of a weak solution, not weak enough to dissolve the albumen off the paper, but of the minimum strength.   The solution, as before, would immediately form the silver chloride, but before the albumen had coagulated at the surface, the solution would penetrate to the interior of the film, and then the formation of the albuminate would proceed nearly equally throughout the whole of the interior.   Evidentiy, then, in this case, the contact of the silver solution would be less prolonged than in the former case.   If the floating be prolonged the silver solution in the interior will become weakened, and par-

tially dissolve the albumen and be carried by the water into
the interior of the paper; it will also partially dissolve off the
surface, and a negative printed on such a paper would have all
the appearance of being dead in lustre, and existing in the paper
itself instead of on the surface.

We may thus summarize :—

1. A paper floated on a strong solution may require long
floating.

2. A paper floated on a weak solution requires short floating.

3. And the strength of the solution may be between the
30 grains and 80 grains to the ounce of silver according to the
amount of soluble chloride dissolved in the albumen on the paper
when the negative is really good as regards opacity and delicacy.

The knowledge of the amount of chloride in the paper
supplied by dealers has to be arrived at somehow, and the
following method will answer. Cut up a quarter sheet of the
paper into small pieces, and place them in a couple of
ounces of methylated spirit. This will dissolve out most of the
chloride, and should be decanted off. Another two ounces of
spirit should be added to the paper, and, after thoroughly soaking,
should be decanted off, and added to the other spirit. The
spirit containing the chloride may then be placed in a glass
vessel standing in hot water, when it will evaporate and leave
the chloride behind. It may be weighed ; but since it is better
to know how much silver chloride (AgCl) would be formed, the
residue should be dissolved in a few drops of water, and a little
silver nitrate added. The silver chloride will be precipitated,
and should be carefully washed with water, and then be filtered,
the paper being opened out and dried before the fire on filter
paper. The chloride is then detached and weighed; 3½ grains
of silver chloride would show that a weak bath should be used,
whilst 10 grains would show that a strong bath was required.

With most brands of albumenized paper directions are issued
as to the best strength of silver nitrate solution for sensit'zing,

and a fair estimate of the chloride present can be gained from such directions.

A weak solution loses much of its strength by each sheet of paper floated, much more proportionally, in fact, than a strong solution, since the same amount of fluid is absorbed by the paper in each case, whilst the amount of silver abstracted from the *whole* is also equal, which reduces the strength per ounce more with the former than with the latter. A weak sensitizing solution, therefore, requires much more attention than a strong one: crystals of silver nitrate must be constantly added to the former. In practice and for general work, then, we recommend a moderately strong bath, the method of making up of which we shall describe.

To make up 2 pints of solution with a strength of 50 grains to the ounce, we shall require 2,000 grains of silver nitrate. This is carefully weighed out in the scales, a piece of *filter paper being placed in each pan.* By adopting this plan freedom from all impurities that may cling to the pans will be avoided, and the silver nitrate will be perfectly pure. Place the silver salt in a large clean bottle, and add half-a-pint of water to it, and shake it to dissolve it. The best water for the purpose is distilled water; but filtered rain, pure spring, or river water answers well. If the water contain any chlorides, it will be shown by a milkiness due to a formation of silver chloride. This must be filtered out when the remaining pint and a-half of water is added. The solution is now ready for use, and, being of the simplest character, is not to be excelled, though the addition of some soluble salts may be advantageous, particularly in dry climates or in very dry weather. Such salts are found in sodium nitrate, or ammonium nitrate, as much as equal weights of either of these substances being added. Thus our formula would stand as follows were these additions made:—

Original Solution.

1.—Silver nitrate ... ... 50 grains
Water .. .. 1 ounce

c

Modified Solution.

2.—Silver nitrate   ...   .   50 grains
     Ammonium nitrate or sodium nitrate   50   ,,
     Water   ...   .   1 ounce

The reason of the addition of the ammonium or sodium nitrate is that prints are better obtained on paper which is not absolutely free from water. When very dry, the liberated chlorine (see page 32) is apt to attack the albuminate, whereas it is deprived of much of its activity when it is able to be absorbed by water, which, in the presence of light, is decomposed into hydrochloric acid and oxygen.*

Hydrochloric acid can attack the silver nitrate present in the pores of the paper, and produce fresh silver chloride. If the paper were quite dry, the liberated chlorine would scarcely be able to attack even the silver. Moisture, though very little, is desirable. In the excessively dry climate of India, &c., in the summer, one or other of these deliquescent salts should be invariably present for the purpose indicated, unless fuming be resorted to.

The sensitizing bath should also never be allowed to be acid with nitric acid, since the resulting prints would invariably be poor.

The best way of securing this neutral state is by keeping a little carbonate of silver at the bottom of the bottle in which the solution is kept. A few drops of a solution of sodium carbonate added to the bottle over-night will secure this. The reason why nitric acid is to be avoided is shown by placing a print in dilute nitric acid. It is well known that darkened silver chloride is unaffected by it; but the print will be found to change colour, and to become duller and redder than if washed

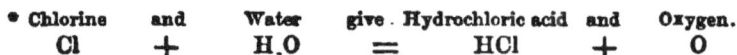

---

* Chlorine   and   Water   give Hydrochloric acid and   Oxygen.
   Cl   + H_2O =   HCl   +   O

in water alone. The nitric acid evidently attacks the albumen. Nitric acid decomposes the carbonate of silver (which, be it remembered, is an insoluble body), forming silver nitrate, and liberating carbonic acid.*

Alum in the printing bath has also been recommended for preventing the bath from discolouring, and it is effective in that it hardens the surface of the albumen; but the ordinary explanation of its effect is defective. If a solution of common alum be added to the silver nitrate we get silver sulphate (which is best out of the bath, and it is slightly soluble in the solution), and aluminium nitrate is formed.†

The same effect would be produced if aluminium nitrate were added to the bath solution. We, however, give a means of adding it as recommended by some writers. When filtering the solution, put a small lump of alum in the filter paper, and pour the solution over it, or add one grain of alum to every ounce of solution, and then filter.

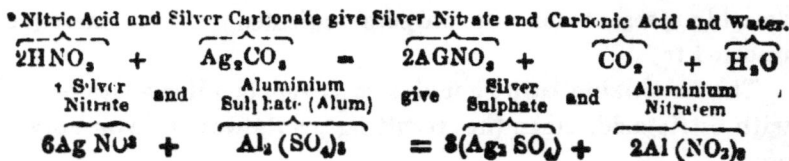

----

* Nitric Acid and Silver Carbonate give Silver Nitrate and Carbonic Acid and Water.

$$2HNO_3 + Ag_2CO_3 = 2AGNO_3 + CO_2 + H_2O$$

† Silver Nitrate and Aluminium Sulphate (Alum) give Silver Sulphate and Aluminium Nitrate.

$$6Ag NO_3 + Al_2(SO_4)_3 = 3(Ag_2 SO_4) + 2Al(NO_2)_3$$

CHAPTER IV.

HOW TO KEEP THE SENSITIZING BATH IN ORDER.

EXPERIENCE tells us, however strong we may make the bath solution to coagulate the albumen on the paper, that a certain amount of organic matter will always be carried into it. At first this is not apparent, since it remains colourless in the solution; but after a time, after floating a few sheets of paper, the organic silver compound gradually decomposes, and the solution becomes of a brown or red tint, and if paper were floated on it in this condition there would be a dark surface and uneven sensitizing. It is, therefore, necessary to indicate the various means that may be employed to get rid of this impurity. The earliest, if not one of the best, is by the addition of white China clay, which is known in commerce as kaolin. A teaspoonful is placed in the bottle containing the solution, and well shaken up; the organic matter adheres to it, and precipitates to the bottom, and the liquid can be filtered through filter-paper or washed cotton-wool, when it will be found decolourized. Another mode of getting the liquid out of the bottle is to syphon it off by any syphon arrangement, and this prevents a waste in the solution from the absorption of the filtering medium. The accompanying arrangement (fig. 4) will be found useful for the purpose, and can be applied to other solutions where decantation is necessary. A is a wide-mouthed bottle holding the solution. B is a cork fitting the mouth, in which two holes have been bored to fit the two tubes, D and C, which are bent to the form shown. When the

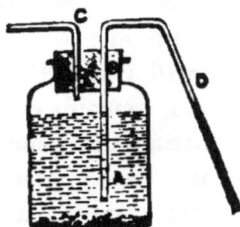

Fig. 4.

knolin has subsided to the bottom, air is forced by the mouth into the bottle through C, the liquid rises over the bend of the tube D, and syphons off to the level of the bottom of the tube inserted into the liquid, provided the end of D, outside the bottle, comes below it.

To bend a tube, a common gas flame is superior to a Bunsen burner. The tube is placed in the bright part of the flame in the position shown; by this means a good length of it gets heated,

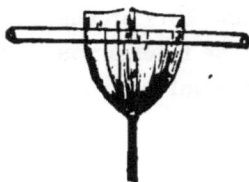

Fig. 5.

and a gentle bend is made without choking the bore, which would be the case were a point of a flame used.

Another method of purifying the solution is by adding a few drops of hydrochloric (muriatic) acid to it. Chloride of silver is formed, and when well shaken up, carries down with it most of the organic matter, but leaves the bath acid from the formation of nitric acid.* This must be neutralized unless a little silver carbonate is left at the bottom of the bottle as described at page 20. A camphor solution may also be added for the same purpose.. Make a.saturated solution of camphor in spirits of wine; and add a couple of drachms to the solution, and shake well up. The camphor will collect the albumen, and it can be filtered out. In case the first dose does not decolourize it, another one must be added.

Another plan is to add potassium permanganate (permanganate of potash) to it, till such time as the solution takes a faint permanent rose tint. The theory is that the organic matter is oxidized by the oxygen liberated from the permanganate, and falls to the bottom. It is not strictly true, however, and the solution will never be as free from organic matter as when the other methods are employed.

The final and best method is to add a small quantity of sodium carbonate (say 5 grains), and expose it to daylight. When the organic matter becomes oxidized at the expense of the silver nitrate, the metallic silver with the oxidized organic matter will fall to the bottom. This plan answers admirably when time is no object, but in dull weather the action is slow. When once the precipitation fairly commences it goes on quickly, and if a little freshly precipitated metallic silver be left at bottom of the bottle the action is much more rapid. This is a wrinkle worth remembering in all photographic operations where precipitation is resorted to.

We have hitherto supposed that the only contamination of the bath is organic matter, but it must be borne in mind that each sheet of paper floated on the solution transfers a certain

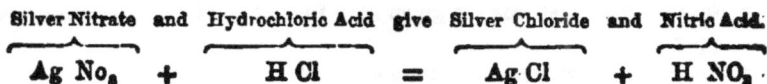

Silver Nitrate and Hydrochloric Acid give Silver Chloride and Nitric Acid.

$$Ag\,No_3 \quad + \quad H\,Cl \quad = \quad Ag\,Cl \quad + \quad H\,NO_3$$

amount of nitrate of the alkali* with which the albumen is salted.

It will thus be seen that in an old bath there will be no need to add the soluble nitrates given in page 17, since they will be already formed. When they are in excess the best plan is to precipitate the silver by some means,† but we select one which is easy of application, since it requires no watching. Evaporate the solution to half its bulk, and slightly acidify it with nitric acid (10 drops to the pint of solution will suffice); throw some ordinary granulated zinc into the jar or bottle containing it; the silver will now be rapidly thrown down in the metallic state, and in the course of two or three hours the action will be complete. Next carefully pour off all the fluid as close as possible to the residue. Pick out all the lumps of zinc, and add a little dilute hydrochloric acid to dissolve up all the small particles of zinc which may be amongst the precipitated silver. Filter the solution away, and wash the residue once or twice with water. Take out the filter paper, and dry it before a fire, or in an oven, and then detach the silver, and transfer it to a small crucible, which place, with its contents, over a Bunsen burner or spirit lamp flame till it is red hot. The heat will destroy all organic matter, leaving a residue of carbonous matter behind, which, after subsequent operations, will be eliminated by filtration. Next cover the silver with nitric acid,‡ and in an evaporating dish slightly warm it over a spirit lamp or Bunsen burner. Red fumes will appear, and when all action has ceased, more acid must be added till such a time

---

* Suppose it is salted with ammonium chloride, we have—

Ammonium Chloride and Silver Nitrate give Ammonium Nitrate Silver Chloride

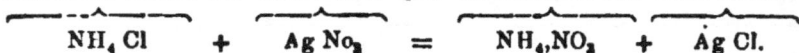

$$NH_4Cl + AgNo_3 = NH_4,NO_3 + AgCl.$$

† Several other methods are given in " Instruction in Photography," in the Appendix.

‡ One part of nitric acid to 4 parts of water.

that very nearly (but not quite) all the silver is dissolved up. Then evaporate off all the fluid and allow it to cool, when water can be added to such an extent that it is *over strength* for the bath. Now measure the whole bulk of the solution in a glass measure, and test by the argentometer for strength. An argentometer is, in reality, an instrument for taking the specific gravity of a liquid. It is as shown in the figure. A B is a glass tube,

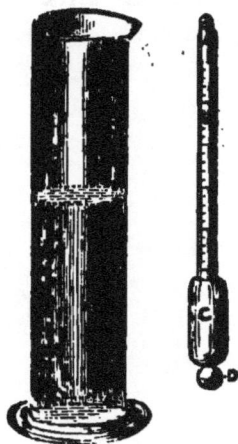

Fig. 6.

inside of which is a graduated scale showing grains; C is a hollow glass cylinder, which has a little glass ball filled with mercury. When immersed in water, the instrument sinks till the scale reads 0—that is, A B is deeply immersed. When any soluble salt is dissolved in the water, the stem rises further. If the soluble salt be silver nitrate, the scale is made to read grains per ounce. It is then evident, if the bath contains any other soluble salt besides the nitrate of silver, the readings will be untrustworthy. Supposing you have a total quantity of 10¼ ounces of solution, and the argentometer tells you it is of a strength of 105 grains to the ounce, you must make a small calculation to see how much water you must add. In 10¼ ounces of solution there will be 10¼ × 105 or 1076¼ grains of silver nitrate. If you want to make the bath 40 grains to the ounce, you must divide this quantity by 40, which is very nearly 27. The original amount of fluid (10¼ ounces), when deducted from this number of (27) ounces, will give you the amount (16¾ ounces) of water that is to be added to give you a bath of the required strength. When the water is added, the solution should be filtered from the carbonaceous matter, and the bath, after neutralizing with sodium carbonate, will be ready for use.

CHAPTER V.

APPLYING THE SILVERING SOLUTION TO THE ALBUMENIZED PAPER.

As each piece of paper takes somewhere about five minutes to sensitize and hang up to dry, it is evident that the larger the piece of paper sensitised the greater will be the saving in time in this operation. Practically a whole sheet of paper, which is about 22 inches by 18, is the maximum ordinary size, whilst it may be convenient to float a piece as small as $3\frac{1}{4}$ by $4\frac{1}{4}$. There is not much difficulty in floating either one or the other if ordinary care be taken, but it is no use disguising the fact that large sheets are sometimes faultily sensitized even by experienced hands, if the solution be not in a proper state. The great enemy to success is the formation of bubbles on the surface of the solution, and if it be at all contaminated with organic matter they are more liable to be met with than if the bath be new. It may be taken as a maxim that no paper should be floated if, to commence with, the bath be not purified. A flat dish of about $2\frac{1}{2}$ inches in height, and an inch larger in breadth and length than the paper to be floated, is used, and the solution poured in to a depth of $\frac{1}{2}$ inch. The paper is grasped by the two hands as shown at page 10, so that a convex albumen surface is formed downwards, which is placed diagonally across the dish and lowered on to the surface of the solution; the hands are at the same time separated outwards, so that the whole surface of the paper is caused to float on it without any arrest. By this means all air is forced out before the paper, and no bubbles should be beneath. To make assurance double sure, the paper is raised from the corners which were not grasped by the hands, and if by any chance a small bubble should be found, it is immediately broken by the point of a clean quill pen or glass rod. Before floating the paper the surface of the solution should be examined for scum or bubbles, both of which may be removed by passing a strip of clean blotting-paper across it. The dish employed should be scrupulously clean, and in cold weather it is a good plan to warm both it and the solution before the fire previous to use. In warm weather, the albumen of the paper may be in a very horny condition. which increases the liability to form bubbles. The writers have found that if the sheet of paper be exposed to the steam passing from a kettle of boiling water for a few seconds (moving it so that every portion shall come in contact with it) just before sensitising, the surface becomes more tractable, and in a better condition for sensitizing; keeping the paper in a moist atmosphere effects the same end.

The length of time for floating the paper depends on the subjects to be printed, *but, as a rule*, three minutes with the 50-grain bath will be found to answer for the majority of negatives. When the proper time has elapsed, a corner of the paper is raised from the solution by means of a glass rod, and grasped by the thumb and forefinger of the right hand. It is then raised *very slowly* from off the solution till another corner is clear, when that is grasped by the forefinger and thumb of the left hand; and it is finally withdrawn entirely, and drained a minute from the lowest corner into the dish. It is next hung up to dry by a corner which should be fastened to an American clip (fig. 7) suspended from a line stretched across the dark room, taking care to keep the corner which last left the solution the lowest. A piece of *clean* blotting-paper about one inch long by ½ an inch wide is brought in contact with this latter corner, and adheres to it from the moisture. This collects the draining from the paper whilst drying, and pre-

Fig. 7.

vents a loss of silver, since it can subsequently be detached and placed amongst the residues for burning.

Fig. 8.

There is another mode of floating large sheets of paper, which is sometimes recommended. One corner is turned up about a quarter of an inch. This is held by the forefinger and thumb of the left hand, and the opposite corner of the diagonal held by the right hand. The first corner is brought on the solution near the oppo-site corner of the dish to that towards which it will eventually be near. The sheet, having assumed a convex form, is drawn by the left hand across the dish, the right hand being gradually turned to allow the whole surface to come slowly in contact with the solution. Air-bubbles are said to be avoided by this means, though for our own part we see no practical advantage in it over the last method.

Some operators also, when lifting the paper from the dish, pass it over a glass rod placed as in the figure, in order to get rid of

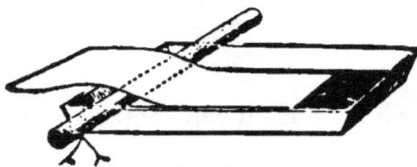

Fig. 9.

all superfluous fluid from the surface. This is a poor substitute for withdrawing the paper slowly from the dish, since capillary attraction is much more effective and even in its action than this rude mechanical means. By those who do not possess patience, however, it may be tried. Some practical photographers also " blot off " the excess of silver, but this is a dangerous practice unless there is a certainty that no " anti-chlor " has been used in preparing the blotting-paper. For our own part we recommend the usual mode of draining the paper. When surface dry, it can be dried in a drying box. The following is a kind which has been adopted by one eminent photographer, and is excellent in principle.

Over a flat and closed galvanized iron bath erect a cupboard. Fig. 10 gives the elevation, and fig 11 the section. A is the bath, D the cupboard, which may conveniently be closed with a roller shutter,* B, passing over *c c*, and is weighted by a bar of lead,

---

* The shutter may be made of American leather, covered over with one quarter-inch strips of oak or well-seasoned pine. The shutter should fit into a groove formed along the sides and bottom of the front of the cupboard.

so as to nearly balance the weight of the shutter when closed. A couple of Bunsen gas-burners, E E, heat the water in A ; the steam generated is carried up the flue F, which also carries off

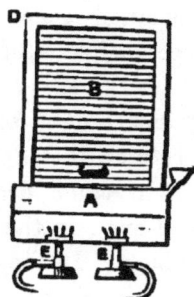

Fig. 10.

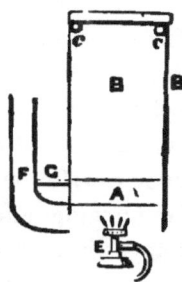

Fig. 11.

the products of the combustion of the gas. The paper may be suspended from laths tacked at the top of the cupboard by means of American clips.

CHAPTER VI.

WASHED SENSITIVE PAPER.

For some classes of work sensitized paper may be washed with advantage previous to drying, and there is much economy in this plan, particularly in hot weather, since it keeps of a purer white for a much longer period than where the silver nitrate is allowed to dry on the surface. It may not be out of place to call attention to the action of silver nitrate on the paper. If a stick of lunar caustic be applied to the skin when dried, there is a peculiar burning effect produced, and even in the dark the cuticle becomes discoloured, though not black. In the albumenized paper we have albumen and the gelatine sizing, and these substances behave somewhat like the skin. The gelatine particularly will become oxidized at the expense of the silver, a reddish organic oxide being formed; and again, if the silver nitrate be alkaline or strictly neutral, we have the same action occurring as when we precipitate metallic silver by means of an alkali, and an organic body such as sugar of milk. The gelatine takes the place of the latter. When the free silver nitrate is removed, the tendency for the spontaneous darkening of the paper is much diminished, since the chloride and albuminate of silver are much less readily reduced than the nitrate. The following plan is adopted for washing the paper:—The paper, after floating, is drawn twice rapidly through a dish of rain or distilled water,

and, unless some other substance which can absorb chlorine be
added to the last wash water, care should be taken not to soak out
all the free nitrate, as then the paper would produce flat prints.
It is then hung up to dry as before.   Immediately before use it
must be fumed with ammonia, in order that the prints may be
" plucky," and free from that peculiar speckiness of surface
which is known to the silver printer as "measles."   We can
readily trace the " measles " to their source.   Suppose all free
silver nitrate is washed away, and the paper be then exposed to
light, the chloride is rapidly converted into subchloride, and chlorine is given off (see page 5) ; if there be nothing to absorb it at
once it will attack the albuminate, which is blackened at the same
time, and fresh chloride will be formed in little minute spots.
These discolour, and are of different tint to the rest of the print,
and give rise to the appearance of measles.   This, of cours , is not
so marked when a little free silver nitrate is left in the paper ;
but as what is removed is principally removed from the surface, it
may still be unpleasantly discernible.   Fuming obviates it entirely
if properly performed, for chlorine and ammonia combine to form
finally ammonium chloride, a neutral and inactive salt.

   Any other chlorine absorber may be substituted ; thus citric
acid, potassium nitrite, and many others are effective, and cause
vigorous prints to be produced.   Perhaps the easiest way of
giving the paper the necessary amount of ammonia is that
recommended by Colonel Wortley.   This is to place overnight
the pads of the printing-frame, if they be of felt, into
a closed box in which is placed a saucer containing a couple of
drachms of liquor ammoniæ, and to withdraw them as required for
the printing-frames.   The pads will be thoroughly impregnated
with the vapour of ammonia, and a couple or more prints,
in succession, may be made before it is necessary to change
them.

   The ordinary method of fuming is that used in America.
Hearn describes a box, which is very convenient and simple in

construction. He says: "Take any common wooden box, large enough for the purpose, and make a door of suitable size for it, which, when shut, will totally exclude all light. Make a false bottom in this about six inches, or so, from the real one, and perforate it with holes of about the same size that a gimlet would make. These holes should be very numerous, and at the centre there should be, if anything, a smaller number of them, because the saucer containing the liquor ammonia is generally placed at the centre of the real bottom of the box."

For our own part we dislike the false bottom as constructed, and recommend one of fine gauze, and, instead of placing half-an-ounce of ammonia in the saucer as Hearn directs, we prefer to soak half-a-dozen sheets of blotting-paper in ammonium chloride solution, about 20 grains to the ounce, and the same number of sheets soaked in lime water; one sheet of each are placed together, and ammonia is liberated by double decomposition; calcium chloride being also formed.

This method is excellent in hot, dry weather, since it imparts a certain amount of moisture to the paper. In damp weather it is a good plan to dry the vapour by sprinkling on the gauze calcium chloride, which will rapidly absorb the aqueous vapour, and will allow the ammonia to pass on unimpeded. The sheets of paper are held at the top of the box by American clips, suspended from laths about three inches apart, and it is not a bad plan to fasten a lath on to their bottom edge by the same means, to do away with their curling. To fume a single piece of paper it may be pinned up to the inside of the top of the lid of a box, and a drachm of ammonia sprinkled on cotton wool distributed at the bottom. The point to be attended to is that the fuming shall be even, and it is evident that the ammonia should rise equally from any part of the bottom of the box. In the plan of the box given above, the bottom of the sheet is apt to get a little more ammonia than the top. The time of fuming depends on so many things that a rule can scarcely be given

D

for it; twenty minutes may be considered about the extreme limit.

If this sensitizing bath be acid, the time must evidently be longer than when it is strictly neutral or slightly alkaline ; and if the negative be hard, it will require to be less fumed than if it be of a weak nature, since ·the action of ammonia is to cause rapid darkening in the deep shadows.   In hot weather the fuming should be shorter than in cold, since the ammonia volatilizes much more rapidly when the temperature is high.   On the whole, we recommend Colonel Wortley's plan of fuming the pads in preference to fuming the paper.

Another mode of preserving the paper from discolouration is to add citric acid to the printing bath, which is effective owing to the fact given at page 32.   The following formula is a good one, and has answered with the writer.   It is—

Silver nitrate ..:   ...   .   50 grains
Citric acid   20   ,,
Water ...   .   1 ounce

The paper is floated for the ordinary length of time, when it is dried thoroughly and placed between sheets of pure blotting-paper.   It will keep in its pristine state for months, if excluded from the air.   It is better to fume this paper strongly 'before use, or the toning becomes a difficult matter.

Ordinary sensitized paper may be preserved for a considerable time if, when dry, it is placed between sheets of blotting-paper saturated with a solution of carbonate of soda, and dried.

Washed sensitized paper is also improved in sensitiveness by floating it for a few seconds on—

Citric acid   ...   .   10 grains
Potassium nitrite   10   ,,
Water ...   .   1 ounce

It can be fumed, when dried, in the usual manner.

In the YEAR-BOOK OF PHOTOGRAPHY for 1880 Mr. A. Borland recommends the following modification :—

He floats the paper on nitrate of silver, as usual, and after it has drained surface dry, blots off any drops that may remain at the edges, and then floats the *back* of the paper for about three minutes on the following bath :—

Nitrate of soda...	1 ounce
Distilled water...	16 ounces

This is rendered slightly acid by a little solution of *freshly* prepared citric acid in water. The degree of acidity is regulated by litmus paper (the blue specimen), which should be slightly reddened by it. After this solution has been mixed about ten minutes, it is filtered, and the paper floated. Mr. Borland says the paper keeps well, and prints the same as ordinary paper, and any tone may be produced.

CHAPTER VII.

CUTTING PAPER.

WE have often come across operators who have no really definite plan on which they cut up their paper for a day's work, and they have little idea of the most economical place of dividing the sheets. The following remarks by Mr. Hearn, which appeared in the PHOTOGRAPHIC NEWS, 1874, will be useful to the printer, and, being so extremely well described, we take the liberty of reproducing them.

"In cutting up the paper for printing, due regard should be given to the materials employed. In the first place, the fingers should be free from anything that will stain or soil the paper, and they should never touch the *silvered* side, but always the *back*. The hands should be perfectly dry, free even from any perspiration, for if this is not strictly regarded in the handling of the paper, 'finger stains' will appear on those parts of the paper with which the fingers come in contact. To guard against this, a rough towel should be suspended in a convenient place, and the hands wiped upon it as often as may be found necessary—say once in every five or ten minutes. An ivory newspaper cutter, about eight inches long and an inch wide, together with a suitable sized pair of shears, will be all of the instruments necessary.

"In cutting the paper for very large prints, such as 13 by 16, 14 by 18, 16 by 20, &c., the beginner had best (to obtain the right size) lay over the sensitive paper the proper sized mat that is to be placed over the print when finished, and then cut accordingly. Considerable paper can be saved in this way, and printed in card size.

" There should always be an assortment of different sized mats in the printing room ; one of each size will do, which should be kept expressly for this purpose.

"In cutting the paper for an 11 by 14 print, the length of the sheet is generally placed before the printer, and the paper bent over to the further edge of the sheet, and then creased, and thus cut into two equal pieces, one of which can be used for the contemplated print. I would recommend that instead of taking exactly one half of the sheet of paper, as described above, to take about *an inch more* than the half, so as to allow for any slight tear that may happen along the edges of the paper during the washing, toning, &c., and also so as to be sure of having the paper wide enough for the different sized mats.

" I have seen some nice prints printed upon the exact half of a sheet of paper, which, when taken from the final washing (and the edges trimmed, being slightly torn), were then too narrow to be covered with the proper sized mats, and had to be rejected ; whereas, if in cutting this paper allowance had been made for this final trimming, the prints would have been saved. The rest of the sheet can be cut very well into sixteen or eighteen carte pieces.

" In cutting cabinets out of a sheet, fifteen is all that can very well be obtained, and to get that number lay the sheet on a wide table, or printing bench (with the length of it running from right to left), and divide it into three equal parts. By laying the cabinet glass on these strips of paper, and cutting the paper a little wider than the glass, five cabinets can be obtained from each strip, and fifteen out of the whole. These pieces will be plenty large enough, both in length and width; besides, this is a very convenient and economical way to cut the paper without waste.

"By a glance at the cut (fig. 12) it will be seen that the size of the pieces will be 4½ by 6 inches, and consequently there will be more room for the width than there will be for the length. The edges of the width side of the paper can be trimmed a little,

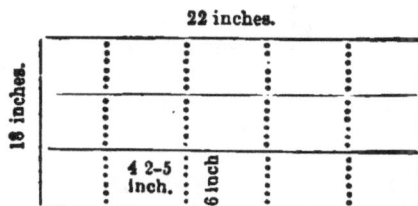

Fig. 12.

as there is usually some little tear, or some other defect, that can thus advantageously be got rid of. Often, when there are only a few cabinets to be printed, I take a quarter-sheet, and bend over the length of it to about three-quarters of an inch of the opposite side, crease it, and then cut with the paper-knife. You thus obtain a large and small piece; the smaller one of these can be cut into four cards, and the larger one can be cut in two, and thus obtain two generous size cabinets; or the printer can use the larger of the two pieces for printing the 4 by 4 size. This is the way I obtain my 4 by 4 pieces when I wish them.

"The beginner must remember that in bending over the *length* of a sheet of paper 18 by 22 inches in size, the divided paper will be 11 by 18 inches in size, which is termed, in the language of the printing room, half-sheet.

"To obtain the quarter-sheet, the *length of the half-sheet* is cut equally in two pieces, and then the size will be 9 by 11 inches.

"A glance at fig. 13 will show that either a generous size 4 by 4, or a couple of nice cabinet pieces, together with four cartes, can be easily obtained from a quarter-sheet.

"To obtain thirty-two cartes, quarter the sheet, and divide each quarter into eight equal pieces.

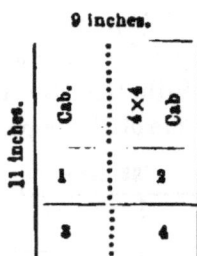

Fig. 13.

"To obtain thirty-six pieces out of a sheet, it is necessary, for convenience, to first quarter it, and then divide it into three equal strips (fig. 14) taken from the *length of* the paper. The

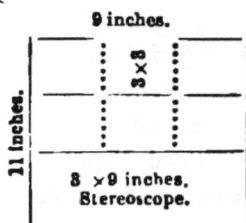

Fig. 14.

pieces, as thus cut, will measure 3⅔ by 9 inches, which will answer admirably for the stereoscopic size. Each one of these strips of paper can be cut into three good sized cartes, making nine out of a quarter, and thirty-six out of a whole sheet.

"Forty-two cartes can be obtained very neatly by laying the sheet before you (fig. 4), and dividing the length into seven equal parts; when done, each strip should measure 3⅐ by 18 inches in size. The whole number of pieces will be forty-two. It will be seen that the size of the carte pieces (3 by 3⅐ inches) only allows very little room for waste paper in trimming after

printing, and thus it will be found necessary to exercise some
care in placing these pieces on the negative for printing.

8 inches.

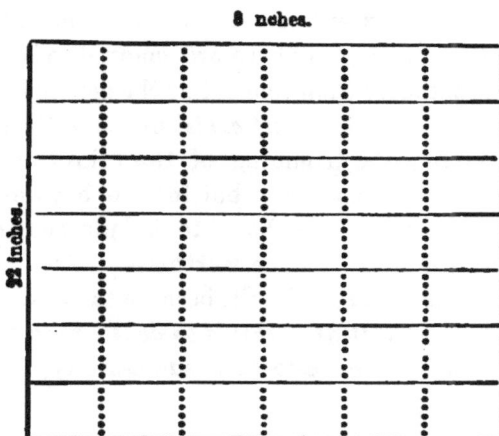

Fig. 15.

" To obtain the forty-two carte pieces from the sheet without
waste, great care is required in sensitizing the paper to prevent
tearing, and also to prevent silver from getting on the back of
it; in cutting it either the shears or the paper-knife should be
used with care. *Do not tear the paper with the hands,* as is very
often done, especially when the printer is in a hurry.

" In making out the above, I have considered the sheet of
paper to be 18 by 22 inches in size, but it is seldom that the
sheet measures *exactly* this, for the *length* often measures from
one quarter to one inch more, but never less, while the width is
invariably the same. When this is the case, a little better margin
is allowed in cutting the sheet up, which is a good thing, especially when a large number of small pieces are to be obtained
from the sheet. Forty-two pieces is all that should be obtained
from a sheet of paper which measures 18 by 22 (or 18 by 22¼,
&c.) inches, because the pieces of paper are now as small as they

should be with safe results to the prints, on account of bad edges, &c., which it is often necessary to trim after printing. There is a way to obtain forty-eight, and even fifty-two pieces of paper from the sheet, but I would not advise any of my readers to try to obtain that quantity, as there are many disadvantages connected with it that more than neutralize the benefits. The paper is sometimes cut up to the *exact carte size*, and then printed up as it is, thus saving the trimming of the prints after printing. This is, perhaps (?), a good way, but for the beginner it is very risky, because the paper will have to be placed *exactly on the negative*, or else the print will be worthless. Even to the experienced printer this is very difficult, because the greatest care and skill are required to do it *as it should be done;* then the *inexperienced* printer could not hope to do it successfully.

CHAPTER VIII.

PRINTING-FRAMES.

There are a variety of printing-frames in the market, each of which may have something to recommend it; and yet, as a rule, the simpler and more uniform the frames are, the more handy are they for the printer, since he rapidly becomes accustomed to handling them, and knows their peculiarities. The simplest pattern is one introduced by Meagher, as shown in

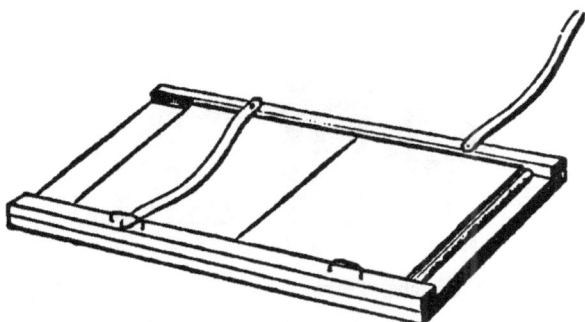

Fig. 16.

fig 16. The negative rests on india-rubber strips which line a framework of its exact size, and a folding back, as shown, covers it. The paper is pressed on to the negative by a pad, and the back on that by means of two brass springs. This is a very excellent pattern for cabinet pictures and cartes, but we can scarcely recommend it for anything larger, since even if it were possible to supply sufficient pressure to secure proper contact of the paper, the negative would be in danger of being cracked.

For all sizes above cabinet, the printing frame as given in the figure is the best. The construction will be seen at once.

Fig. 17.

In the front part of the frame is a piece of thick plate glass (depending for its thickness on the size of the frame). On this the negative rests, and over this again are the necessary pads and backboard, which is clamped down by means of two cross-bars, on which springs are fixed. An increase of pressure may be given by increasing the thickness of the pad (which may consist of smooth felt) next the negative, or by sheets of thick blotting-paper quite free from all folding marks.

Sometimes the back of the frame is hinged in three pieces,* and this is almost essential for large prints (say 2 feet by 1 foot 6 inches), since every part of the picture should be capable of examination during the progress in printing. With a simple single hinged backboard this is impossible.

---

* In fig. 18 the fastening for only one of the pressure-bars is given, to avoid complication.

When large negatives are to be printed, the plate glass
front should always have at least an inch .clear all round.
For smaller negatives (say 12 by 10 and under) half-an-inch

Fig. 14.

clear is sufficient. This allows a certain latitude in the position
of the negative, and enables the fingers to get at the paper

Fig. 15.

without inconvenience. In the frames in which the front of the
negative is unsupported this cannot be the case, and for this
reason (as well as those given above) they are not recommended
for large prints.

CHAPTER IX.

PREPARING A LANDSCAPE NEGATIVE FOR PRINTING.

LANDSCAPE negatives are rarely ever in perfect harmony for printing, and much may be done by judicious doctoring of the best of negatives to secure the best of prints. With moderate negatives it is absolutely essential that they should be improved. Let us take the example of a hard landscape negative, which if printed so that the deep shades should show detail, would show none in the high lights.

A piece of thin tissue paper (the kind known as *papier minerale* is the best), of the size of the negative, is damped evenly with a sponge, and carefully pasted on the back of the negative. The negative is then held up to the light, and the high lights carefully traced with a faint line by means of a pencil. These are then cut out by means of a sharp penknife, and a trial print taken in the shade. If it be found that the shadows still print too deeply when the detail in the high light is visible, another thickness of tissue paper may be applied, cutting out this time, perhaps, the high lights and the half tones. Another trial print will show whether the object is attained. If still not satisfactory, crayon in powder from the scrapings of a stick of crayon, or blacklead, may be applied by a stump to the parts requiring it. It may happen that the effects of the tissue paper may be seen in the print by the light penetrating beneath it, and causing the edges of the shadows to print too dark. In this case, which may arise from the negative being taken on a thin glass plate, the parts covering the high lights, and which were cut out, should be indented with a jagged edge such as this, the dotted

Fig. 16.

line showing where the cut would come if it had been cut out in a clean sharp line. Another mode which we have sometimes found successful, though care is required in employing it, is to coat the back of the plate with a very dilute emulsion of a quarter the ordinary consistency, than to expose it, through the negative, and develop with one of the ordinary alkaline developers (we prefer the ferrous oxalate),* and then fix. This last film may be protected with a layer of albumen 1 part of albumen to 25 parts of water. By this means the shadows become subdued and the contrasts diminished, and there is no danger of any sharp demarcations in the shades being apparent.

There is one way of improving a hard negative, if taken on a gelatine plate, which would probably be dangerous in the hands of a novice, but which is most effective when used with skill and judgment, but must be applied before the plate is varnished. One of the most popular methods of reducing the density of an over-intensified gelatine negative is with a very weak solution of perchloride of iron. The writers have found that the reducing agent may be applied locally. Let us suppose the case of a figure in a landscape in a light dress, which produces a white patch in the print. The negative should be placed in a dish of water, then lifted up until the part to be reduced

---

* See " Instruction in Photography " (page 67), fourth edition.

is just above the level of the water; a solution of perchloride of iron should then be applied to the part with a camel's-hair pencil, care being taken that it does not spread over the edges or run down the negative. When this is found to be taking place, the plate should be allowed to fall into the water; it can then be lifted again, and the operation proceeded with. It is not easy to give any strength for the solution of perchloride of iron, but it is /best to begin weak, and strengthen as required. A saturated solution has been used in an obstinate case without any mischief being done, but this required very careful watching.

For landscapes, Mr. England has successfully used a strong solution of cyanide of potassium with the same object. He moistens the parts of the gelatine plates which require reduction with water applied by a paint-brush, and afterwards, with another, applies the cyanide. The reduction can be watched as it progresses, and by a judicious use of the brush no sharp line of demarcation between the reduced and unaltered parts is visible.

With a thin negative the tissue paper may be applied as before, only in this case the shadows are left bare, the half tones have one thickness of tissue paper over them, the highest lights two or three. An emulsion may be used in this case as well, only instead of fixing the transparency which is at the back, the precipitated silver is dissolved away by nitric acid, and the developer applied again. By this means, the density in the high lights may be doubled if required. It must again be repeated, that in all cases the use of emulsion requires great care, seeing that if any get on the varnished surface, markings are sure to occur. It sometimes happens, especially with gelatine plates, that the corners of one side of a negative print too dark. This is very visible in sky and sea pictures. The careful application of blacklead on the tissue paper on the back of the plate may often save a beautiful negative that would be otherwise useless.

In most landscape negatives there is a want of atmosphere

(by which we mean the haze always present in the air) in the distance and middle distances, and we have found that by applying one piece of tissue-paper to the back of the negative to cover the middle distance and distances, and another to cover the distance alone, atmospheric effect is produced.   The effect of atmosphere is usually shown by grey tones as compared with those of the foreground, and the greyer they are the more distant should the objects be away in nature.   This effect is accomplished by the tissue-paper.   It must, however, be remembered that the lights of distant objects are greyer than those of the foreground, hence the tissue paper must be used with judgment to prevent the distant lights from appearing too white.   This sometimes is effected by giving *the lights in the foreground* a covering of tissue paper.   We very much doubt if there exists any landscape negative which would not be improved by the use of tissue paper, since photography often tends to do away with atmosphere. We have, in some cases, strengthened the high lights on the film side with the paint-brush and Prussian blue. This requires skill, and should be done very sparingly.   It may be objected that when these artifices are resorted to, that the photograph must of necessity fail in regard to truthfulness.   The answer to this objection is quite easy to give.   If a photograph were true in itself, they should never be resorted to, but since it always falls short of the truth, it is quite legitimate to give it the effect that a perfect process would do, by which we mean one in which the intensity of the negative is exactly proportional to the intensity of the light producing it.

It has been shown in the PHOTOGRAPHIC NEWS of 1877, that the gradations of a negative are never perfect, and the use of the tissue paper, &c., makes it more nearly in accord with nature.

These remarks, of course, have reference only to what we might call "a good printing negative;" the advisability of doctoring poor negatives is scarcely open to argument.   Improve as much as you like, but be very careful not to overdo it.

CHAPTER X.

PRINTING THE LANDSCAPE PICTURE.

A TRIAL print from a negative should first of all be taken, to enable the operator to guage as to how much is required to be done to it. A piece of sensitized paper of the exact size of the plate is taken and examined by transmitted light in the dark room. If there be any appearance of markings due to bubbles, or of star-like metallic spots, probably due to small particles of iron being in the albumen, it need not be rejected altogether. Should there be any of these defects, the sheet should be placed on one side to cut up into smaller sizes. We will suppose that we are going to print a 15 by 12 negative. A strong frame (of the description given at page 44) must be employed, and the thick plate glass carefully freed from all dust, grit, or stains. The back of the negative is then placed in contact with it, so as to occupy the centre of the frame. The piece of sensitive paper is placed over it, and the back placed loosely over it, and is then carried face downward into the place where the printing is to be done, and the frame is placed face downwards on the floor, and left for a few minutes. By this artifice the paper takes the same degree of humidity as the atmosphere, and there will be no danger of any cockling, and consequent (as it is termed) want of contact, between the paper and the negative. This is

E

only necessary when there is any very great difference in the temperature of the drying room and the place where the prints are to be exposed, and in some establishments the difficulty is met by carrying the whole supply of paper in a closed box into the latter place, and allowing it to absorb any moisture that it can take up. In any case, the paper is next to be placed in absolute contact with the negative, and we strongly recommend the use of sheets of blotting-paper cut to the proper size (about four thicknesses will be sufficient), and backed by a thick pad of closely woven and very smooth felt. These latter are rather expensive, but are very durable if ordinary care be taken of them. The blotting-paper is useful in causing contact, and also because any accidental presence of silver nitrate solution on the back of the sensitive paper is immediately detected, and there is, consequently, no danger of carrying it to another print and spoiling it, which it might do were it absorbed by the felt pad.

The back of the frame is then placed *in situ*, and the hinged cross-piece brought down and secured by the fasteners. If the springs be sufficiently strong, the film of the negative should now be in absolute contact with the sensitive paper. If there be any grit on the plate glass, or adhering to the back of the negative, it is highly probable that the glass plate will crack, and if the plate on which the negative is taken be very curved,* the same disaster may be expected. Suppose the day to be bright, and the negative fairly dense, the frame may be placed for the trial print facing away from the sun (if there be any) so that it receives merely skylight, and no direct rays. When the transparent parts of the negative seem to have taken a fairly black or brown colour, the print should be examined. In practice we have found (supposing the printing room be away from the dark room) that a cloth of thin yellow calico is a useful

---

* For this reason, amongst others, it is desirable that photographers should use glass for their negatives which is at least tolerably flat.

adjunct during the examination. The cloth is large enough to cover the frame and also the head of the operator. One half of the back is loosened and raised, the half pieces are pulled back, and the paper will probably be found adhering to the negative, and may require a little manœuvring to separate it. A very thin slip (of the size of a toothpick) of soft wood, sharpened at one end, is a good implement to employ, as by inserting it the paper can be separated at one corner, and then be raised by the fingers. We have seen some printers blow against the paper, as if they were separating the leaves of a book from one another, but this method is to be deprecated, since particles of saliva are apt to be carried on to the paper with the breath, and to cause spots, which often appear unaccountable. Should the print appear slightly deeper than it is required to remain, it is probably ready to be withdrawn from the action of light, but the remaining half of the paper must next be examined to see whether such is the case. To do this the first half of the pressure-board of the frame which is loose must be pressed down once more into position, the frame reversed end for end, and the other half of the board opened.

If the print is large (say 15 by 12) it is not advisable to look at much of it at once, or for a longer time than can be avoided. It constantly happens that on a warm day the paper contracts during the short time necessary for a proper examination of the print; the consequence is, that the paper does not fall on the same place on the negative when reflected, and the result is a double print on the paper.

The printing being judged to be complete, the paper is withdrawn by taking off pressure-board and pads, and put away for the further operations of toning and fixing. In one establishment we are acquainted with, the prints when taken from the frame are placed in a box the lid of which is pierced by a hole covered with a dark cloth; whilst others keep them in a press of blotting-paper. The great point to attend to, however, is to keep them away from all actinic light; and we should say, further, from all light, since darkened silver chloride becomes oxidized in light which is usually considered to be non-actinic. No doubt every printer is aware that the prints produced from the same negative and on the same sample of albumenised paper similarly sensitized vary considerably in richness and depth on different days. For instance, when the light is bad, and when, consequently, the printing takes a long time, the colour of the darkened surface will be found to be much duller than on a day when the light is powerful. Silver albuminate is much less sensitive to feeble light, whilst in bright light the difference in sensitiveness is not nearly so marked, and this may account in a certain degree for the difference; but if any one takes the trouble to expose sensitised albumenized paper to bright light so as to darken, and then to cover up half, leaving the other half to be exposed to the light coming through ruby-glass, it will be found that there is a difference in colour between the two portions, and on toning the differences will be still more marked. In dull weather the red and yellow rays bear a greater proportion to the blue and violet rays (all of which enter into the composition of white light) than they do on a bright day. It is the blue and violet rays that reduce the silver chloride to the state of sub-chloride, and then oxidize the latter; yet it must be remembered that the red and yellow also oxidize the sub-chloride without being able primarily to produce it. Hence on a bright day, when the printing is quick, the red and yellow rays have but little time to do any work, whilst on a dull day they have plenty of opportunity of oxidizing the sub-chloride as fast as it is formed. The oxidized image is always more difficult to tone than one which is unoxidized, hence the advantage of printing in a good light if possible. The writers believe that one of the principal causes of the variation in tone of silver prints, which is only too often to be seen, is caused by this difference in length of exposure to the light.

The operator must now be supposed to be cognizant of the operations of toning and fixing which are to be described in subsequent chapters, and that he has the finished trial print of the particular landscape negative before him. He sees whether the middle distance or far distance is obtrusive, and notes which portions require to be softened down by tissue paper, or to be brought nearer by strengthening the high-lights, and eventually forms a picture of it as it should be, centreing his imagination in it as built up round the point of principal interest. He endeavours to see whether the sweeps of light and shade lead up to this principal object in the view, and whether, if light, it is in contrast with an immediate dark part of the picture, or *vice versa.*

Knowing that this is one of the laws of art, he next should endeavour practically to give effect to his imaginative picture by the judicious manipulation of tissue paper, the crayon, and the paint, such as described in Chapter IX. The next point to attend to is as to whether the picture requires clouds or not, and if he have a stock of cloud negatives of the right size, he must endeavour to pick out one, a portion of which will compose well with the lines of the picture,* and at the same time be correct as regards light and shade. When such a negative is selected, it remains to print it in. A white sky is an abomination, and a plain tinted one without gradation is nearly as bad. If, therefore, the operator has the heart and means to do this double printing, he should never neglect to do it.

But we would here remind him that when a sky-negative has been used with a particular view, it should always be devoted to that landscape. Nothing could be in worse taste, or further from nature, than to use the same sky with different landscapes. We once saw a frame of sixteen views, thirteen of which were backed with the same sky; this was bad enough, but the absur-

---

* See " Pictorial Effect in Photography " (Piper and Carter).

dity went further, and in the same exhibition were landscapes by another photographer with the same sky! The inference is that both these photographers bought their sky negatives, printed them, and exhibited them as their own—a proceeding to which a harsh name might be given. To use a cloud negative properly, the reader should consult the chapter on "Combination Printing."

There is another artifice, however, that does away with the blank sky. It is practised by some of the leading photographers in England, and may be put in requisition instead of the more elaborate double printing. In order to do this, a not quite opaque sky—that is, one which "prints in" a little—is necessary. Very effective clouds may be produced by a paint-brush and lamp-black, Indian ink, or gamboge, by painting them artistically *at the back of the negative*. It matters not if the clouds so formed show sharp lines and dots, since, if the printing be done in diffused light, the thickness of the glass plate on which the negative is taken shades these off, and gives them the soft edges which are natural to clouds. The clouds may take any of the usual shapes as seen in nature, and the paint should not be applied too strongly, but should have a certain amount of transparency. In some negatives we have seen taken on dry plates, the sky was very transparent, and, when printed in the ordinary manner, showed a good deal more than perceptible tint. Yet, by a judicious masking, fleecy clouds floating in a light sky were produced, which deceived the greatest connoisseurs in such matters.

We will now describe how such a negative should be prepared for printing.

Black varnish should be carefully run round the sky-line on the face of the negative, for about a quarter of an inch. On the back of the negative the medium should cover the sky to within one-eighth of an inch of the sky-line, and by this means a sharp but *slightly softened edge* of the distant landscape was projected. The breadth of the black varnish border on the back was slightly greater than that on the film side of the negative, being about an inch. A piece of cardboard was also roughly cut out to the sky-line, and left sufficiently broad so as to more than cover the sky when laid flat on it. The negative with the clouds painted on it was now placed in the pressure-frame, with the sensitive paper in contact with it. Outside the frame, and corresponding with the sky-line, the edge of the cardboard was placed, a small bar to act as a weight was placed across it as shown in the figure, and the top end supported by a

Fig. 20.

couple of wooden pegs. The printing took place in diffused light. When the picture was withdrawn from the frame, the sky, being shaded gradually by the card, was printed in lightly, whilst the remaining portion of the negative received the full light; the sky, as is right it should be, was darker near the zenith than toward the horizon, where it was, in fact, white; but since the clouds were printed in at the top, the baldness of the white sky was avoided.

Excellent clouds may also be produced by the stump and crayon on tissue paper, many of the effects of delicate clouds being capable of being produced in this manner. A certain amount of skill is required in producing them, but nothing beyond that which a little practice can give.

We may add that, instead of using this cardboard shade, some printers prefer first to entirely mask the sky and print in the landscape, than to mask the landscape, and to use a movable screen over the negative, drawing it backwards and forwards during exposure, taking the precaution that the top of the sky receives the most exposure. The method of using the cloud negativ, we have already said, will be found in the chapter on "Combination Printing." Above all things, the printer must bear in mind that if there be any *distance* in the picture, the sky, when it meets the margin, must be only very delicately tinted. Let it be remembered that a picture is often spoilt by printing in clouds too heavily. The clouds for an effect should be most delicate, with no heavy massive shadows which overwhelm those of the landscape itself. We are only talking of the ordinary landscape when the effect of storms is not desired. It is not within the scope of this work to show how a landscape and a sky negative may be printed into one plate to form a transparency from which a new negative may be made; suffice it to say that, by using collodio-chloride, or by the use of a slow dry plate and exposing to candle light, the former may be produced in almost the same way that the print is produced, and a negative may then be produced in the camera or by a dry plate.

CHAPTER XI.

So much has been written on the subject of what is called "re-touching" the negative, that it would be a waste of space to enter very fully into details here. It is now generally admitted that working on the negative is not only legitimate, but that it is absolutely necessary, if a presentable portrait is to be printed. The only question is, where to stop. Professional retouchers, in too many cases, do too much, and by doing so they "overstep the modesty of nature," and turn the lovely delicacy, softness, and texture of living nature into the appearance of hard and cold marble statuary. Everything that is necessary to do to a portrait negative is very simple; it should be corrected, not re-modelled. Freckles and accidental spots should be stopped out, high lights may be strengthened, and shadows softened. We may here briefly indicate the technical methods of performing these operations.

Some operators pour a solution of gum over the negative after fixing, and when it is dry work upon the surface of the gum; but it is better and safer to retouch the negative after it has been varnished. The varnish must be allowed to become thoroughly hard before any working upon it is attempted. A negative varnished at night should be ready to be retouched the next morning. If very little has to be done to the negative, it may be done at once without preparation; but it is often advisable to prepare the surface of the varnish to take the lead pencil, with which the greater part of the work is done. This is done with "retouching medium."

Several preparations under this or similar names are sold by stock dealers, all of them giving, as far as we have tried them, equally good results. If the photographer prefers to make his own medium, he may do so by diluting mastic, or any similar varnish, such as copal, with turpentine. Apply the medium to the parts that it is intended to work on with the finger, and allow to dry, which it does in a few minutes. Place the negative on a retouching desk, and commence to fill up with the point of the pencil all spots that are not required, such as freckles or uneven marks. Some operators begin at the top of a face and work evenly downwards. This is a bad plan, and usually results in a mechanical flattening of the face; it is better to fill in here and there as necessity appears to arise. The high lights may now be strengthened, taking care not to make them violent or spotty. The shadows of the face will be found to require softening, but the general shape of the shadows must not be altered, and in modifying lines—such as the lines in the forehead and under the eye—take care not to remove them altogether. An old man without wrinkles is an unnatural and ghastly object—the "marble brow" of the poet should be left to literature. The best pencils to use are Faber's Siberian lead, the hard ones in preference. HH and HHH are the sorts usually employed. The pencils must be kept very finely pointed. To ensure this, a piece of wood covered with glass cloth should be kept always at hand on which to grind the leads to a point.

Sometimes there are portions of a negative that require more filling up than can be done with a pencil; in this case water-colour must be employed. Indigo or Prussian blue is, perhaps, best for the purpose, because these pigments allow a more appreciable or visible quantity to be laid on without becoming opaque than any of the warm colours. Sometimes parts—such as the arm of a child—will print too dark when in contrast with a white dress; in this case it will be necessary to paint over the part on the back of the negative, or to cut out a piece of *papier minerale* to the shape, and paste it over the dark part, also on the glass side of the negative.

CHAPTER XII.

OF the many varieties of small portraiture, the vignette is, perhaps, the most popular, and, when well done, is certainly the most refined and delicate. Two things are to be especially avoided in vignetting. The form of the vignette should *not* follow the form of the figure closely, as it is too often made to do, and dark backgrounds should not be employed. The qualities to endeavour to attain are softness of gradation, and an arrangement of the forms of the vignette that shall throw out the head and figure, and the resulting print should somewhat resemble a sketch, finished if you like, but sketchy in effect. Although the background should be light, it ought not to be white, but of a tint that would just throw up the white of a lady's headdress. If the background screen could be painted so that a little shade should appear over the shoulders of a sitter for a head, or rather darker behind the lower part of a three-quarter figure, so much the better would be the effect. It would be difficult to find a case where gradation could not be of advantage in a background; however slight, it conduces especially to relief.

Having stated what should be aimed at in vignettes, we now come to the technical methods of producing them.

In many cases vignetting is considered to be a merely mecha-

nical operation, and very often looks like it. Perhaps the trade
have more to answer for than the printer, since the qualities of
the wares advertised for the use of the vignetter are often ex-
aggerated to such a degree that they are supposed to be suitable
to any pictures. Vignette glasses are not so common as they used
to be, but they certainly are useful in some instances; we almost
think that the methods of producing vignetting apparatus which
will be described shortly, superior to them. In case the printer
should wish, however, to use these glasses, here is a method by
which he may produce them. Have a piece of orange glass,
flashed on one side only, rather larger than the size of the picture
to be vignetted. Take a rough print, and trace round, in the
proper position on the glass with an ink line, the point to which
the picture should extend. This should be marked on the un-
flashed surface of the glass—that is, the surface on which the glass
is uncoloured. Place the plate so marked on a white surface,
flashed side uppermost, and make a solution of hydrofluoric
acid and water, 1 part of the former to 3 of the latter, in a gutta-
percha dish or bottle.* Make a pad of flannel and cotton wool
at the end of a stick, about the size of a large nut, and drop
this into the solution. Dab this on the coloured surface of the
glass in the central portions where the print is to be completely
printed in, gradually working out to the inked line. Always
work from the centre to the edges, and dilute the acid with a
little water as it approaches the margins. By degrees the
flashing will be dissolved away in the centre, and, if properly
performed, the colour will gradually be eaten away, till the
glass is colourless in the centre, and keeping its full shade of
orange at the ink lines. The glass is then washed, and is ready
for use.

---

* Hydrofluoric acid is always supplied by chemists in gutta-percha
bottles, as it attacks glass. A spare gutta-percha bottle can easily be procured.

The most popular plan of vignetting is with cotton-wool. We believe that the greater part of the vignetting done in England is by this clumsy, costly, and difficult method. It requires more time and attention than any other way of producing the same results. Its advantages are, that it is more " elastic," and allows the operator more scope for attention than other methods. In the hands of a person who has very great skill, taste, and patience, it is undoubtedly most useful ; but when used by anything lower than the highest skill, the results are almost always hard and inartistic. The operation is thus performed. A hole is cut in a piece of cardboard, which is placed over the negative. Under the edges of the cardboard is placed cotton-wool, which is lightly pulled out, so as to slightly shade the vignette, and produce the vignetting gradation.

The next methods of vignetting are dependent on simple laws of optics. Suppose you cut a round hole in a card, say, half-inch in diameter, and so arrange it that all the light getting to a sensitive paper comes through this hole, and that the card is for our experiment placed half-an-inch from the paper. Now place the hole so as to face the sky, but so as the sun has no direct rays falling through the hole. It will be found that the greatest darkening will not occupy a space exactly opposite the hole, but be *away from the side on which the light is brightest*. The dark round patch will be shaded gradually off till a line is reached where, practically, the light has no effect—that is, if the surface of the card next the paper be blackened. It will be noted, however, that the shading is not equal on both sides, but that the gradation is most extended away from the side on which the light is brightest. A good example of what is meant will be to try the experiment of placing the paper and card flat on the ground in the angle between two walls, both of which are in shadow. It will be seen that the brightest gradation takes place in the direction exactly away from the angle of the walls. Next repeat the experiment, making the hole point to the sky, which is equally illumi-

nated and pointing well away from the sun. It will be found that
the gradations are equal, and the greatest darkening exactly
opposite the hole.   Raise the card next to the height of one inch,
and the gradations will be found to be more extended and softer.
The reason of this can be well understood by a glance at the
figures. In both, suppose A B to represent the section of the card,

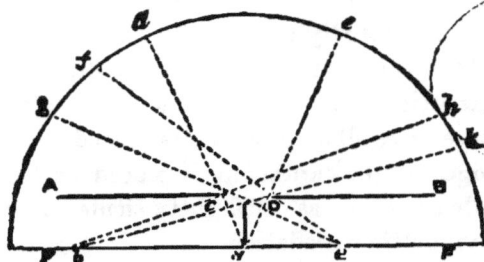

Fig. 21.

and C D the hole in it, and the dotted circle the sky, and E F the
paper.   Take the points *a*, *b*, and *e* on the paper, and let us in the

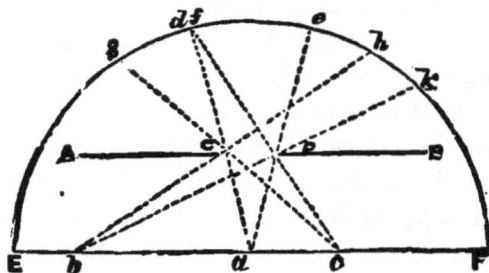

Fig. 22.

three instances see what relative illumination they will receive.
a is opposite the hole, and receives the light from a circle of sky
of which *d e* is diameter, and *b* from an ellipse of which *h k* is
one diameter, and *e* from an ellipse of which *f g* is one diameter.
In the first case, where the card is ⅛ inch from the paper, *h k* is
about one-fifth of *d e*, and *g f* about one-third of *d f*, and since the

ellipses vary as their two diameters multiplied together, the point
b would receive only one-twenty-fifth the light that *a* received,
and *c* about one-ninth.

In fig. 21 the card is raised one inch from the paper, and here
f g is about three-quarters of *d e*, and *h k* about two-fifths;
therefore, in this case, the light on B would be only four
twenty-fifths, or about one-sixth of that acting on *a*, and about
nine-sixteenths or one-half nearly on *c*. It is thus evident that
the further away the card is, the more extended will be the
gradations. Again, suppose, in the last figure, the bit of sky at
g f was twice as bright as at *d e*, then the amount of light acting on *c* would be the same as that acting on *a*. It will thus
be seen how important it is for proper gradation that the hole in
the card should be exposed to an equally illuminated sky, or that
some artifice should be employed to render the illumination equal.
If we paste a bit of tissue paper over C D, this is accomplished,
for then it becomes the source of illumination, and it is illuminated equally all over, since on every part it receives the light of
the whole sky; but this is not the case if it is transparent to
diffused light, and is never the case if it is exposed to direct
sunlight, since a shadow of the hole is always cast on the paper
beneath. If you choose to put another piece of tissue paper,
(say) one inch above the hole, and extending over the whole
length of the card, this difficulty is got rid of, and this last piece
of tissue paper illuminates that pasted over the hole C D, and the
gradations will then be nearly perfect.

Now to apply the above to forming a vignetting block.

Suppose we have a one-inch head to vignette and to show the
shoulders and chest, to be of the size of a carte-de-visite, that the
background is about a half-tone between black and white, and
that but a trace of it shall appear above the head. To make a
good vignette, the graduation from black to perfect white should
lie within a limit of half an inch for a carte size portrait. The
question then arises at what distance from the plate should a

vignetting card be cut to help this object, and what shape should be made the hole in the card. We take it that one-fifth of the light necessary to produce a full black tone would hardly produce any effect on the sensitised paper; knowing this and the size of the aperture, we can calculate exactly what height the card could be raised. Take the breadth between the shoulders that is to be fully printed as 1¼ inches, then by constructing a figure similar to figures 18 or 19 we shall find that the necessary height is about one-third of an inch.*

By judiciously cutting out an aperture in the card and vignetting, defects in a background may often be entirely eliminated from the print. Proceed in this way: Take a print of the portrait, and cut out the figure in such a.way as to get rid of the defective background, and then place this on a piece of thick card (we prefer a thick card, since it will not sag easily, and thus alter the gradation), and cut out an aperature corresponding to it. The outsides of most carte-de-visite frames are raised from the glass about one-third of an inch; place the card on the front so that the

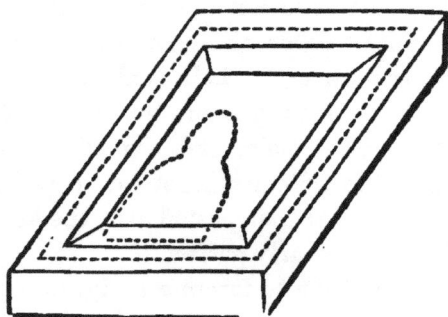

Fig. 23.

aperture corresponds to the figure on the negative, and tack it on to the frame. The dotted lines (fig. 23) show the card

* This calculation is near enough for our purpose. There are certain niceties which might be introduced, such as the "critical angle of the glass."

F

fastened on to the frame, and the opening left. This latter
may be covered with tissue paper, and the frame placed in diffused
light from the sky. In some cases it may be necessary to use a
larger printing frame than the ordinary carte frame, in which case
the operator should be able to make a vignetting apparatus raised
at a proper heigth from the glass. Suppose it is required to raise
the opening half an inch above the glass, and that the card is
4¼ by 3¼.

Take the card and rule rectangles as shown (fig. 24), the inner

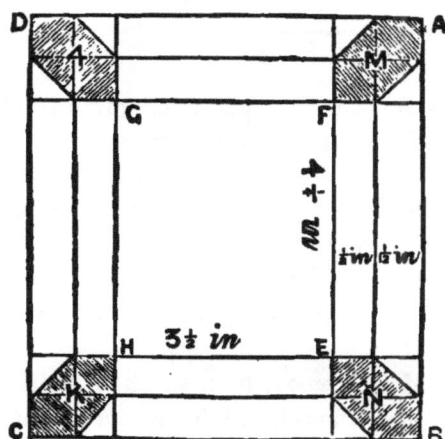

Fig. 24.

one being 4¼ by 3¼ inches, the next one ¼ an inch outside that,
and the third ½ an inch outside that again. Cut out the outer
rectangle entirely, so that we have a piece of card of the size
ABCD. Take a needle point, and prick through the card at the
points EFGH and join these points at the back of the card by
lines. Now take a sharp penknife, and, having laid a flat edge
along, cut the card half through its thickness along KL, LM,
MN, and NK. Turn the card over, and cut along the lines
corresponding to EH, HG, GE, and FE, also half way through
the thickness of the card. Turn the card over once more, and

cut out the shaded pieces at the corners. Now bend the card along the cuts, and a raised block will result of this shape. The

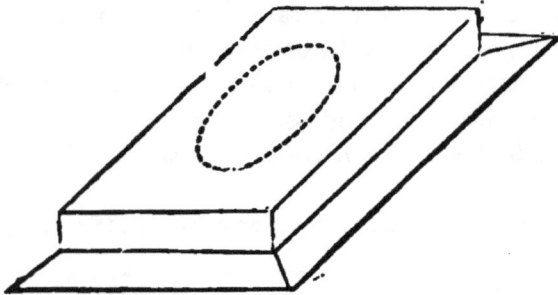

Fig. 2 .

corners are held together by pieces of gummed or albumenized paper, and the block is ready for an aperture to be cut in it according to the portrait to be printed. , Wooden grooves may be glued along the top of the vignetting frame, into which cards containing other apertures can be slipped.*

The most practical method of vignetting, a modification of the above, and the one we always prefer in our own practice, is as follows :—

Take a piece of soft wood, half an inch thick for a cabinet size—a thinner piece should be selected for a smaller picture—of a larger dimension than the negative ; in the centre of this cut a hole of the shape of, but much smaller than, the desired vignette. One side of the hole should be very much bevelled away, as represented in this section (fig. 26). Place this block

Fig. 26.

on the glass of the printing-frame, bevelled side under, the hole

---

* The boxes in which children's puzzles are often packed will give an idea of what is meant.

being exactly over the part of the negative from which the vignette is to be printed. The hole must now be covered with tissue paper or ground glass, and the frame placed flat on a table to print. The size of the hole in relation to the size of the vignette will be easily ascertained by a little experience without the labour of elaborate calculations. On dull days the tissue paper or ground glass may be omitted.

This method is very simple and effective. A quantity of vignetting blocks of various sizes and shapes could be made by a carpenter, or by the printer, and should be always at hand.

A vignetting block should never be less than a quarter of an inch away from the glass, otherwise the gradations will be too abrupt.

CHAPTER XIII.

ARTISTIC METHODS OF PRINTING THE PORTRAIT.

HAVING discribed in the last chapter the various mechanical arrangements by which a simple vignette is produced, we will now proceed to give some account of how that and other forms of printing can be turned to the most artistic account in portraiture.

The idea that printing is a mere mechanical operation was exploded long ago. It is now recognized that the final result owes a good deal of its artistic effect to the way in which the negative is dealt with after it is varnished, and especially to its treatment by the printer. There are many varieties of vignettes, aud the method is useful in various ways.

Plain Vignettes.—The usual vignetted portrait is that which represents a good-sized head and shoulders in the space allotted to the picture. For a carte-de-visite, a head measuring about $1\frac{1}{4}$ inches from the top of the head to the chin is a good proportion. Larger sizes are often made, but they look coarse and vulgar, as if the photographer had tried how much quantity he could give for the money, regardless of quality ; and even if the quality is good, the vulgar effect is still there. For a cabinet size a head of $1\frac{3}{4}$ or $1\frac{1}{4}$ inches is quite large enough. A very pretty style is that in which the gradation is carried out so gradually as only to end with the edge of the paper.

Three-quarter Length Vignettes.—A three-quarter length figure of a lady, either standing or sitting, makes a pretty picture; for gentlemen, a three-quarter vignette is not so good, although it is admissible. It is difficult to make the legs look anything but awkward when they are vignetted into empty space at the knee. For three-quarter vignettes a light, sketchy landscape background may be used with effect.

There are many varieties of what may be called "fancy printing," in which the vignette takes a conspicuous part. The first style that we will consider is that of

Vignettes on a Tinted Ground.—Print a vignette in the usual way. Take it out of the frame, and place it on a board covered with velvet or flannel, to prevent the paper shifting. Cover the print with glass, and place over the printed part a piece of black paper roughly torn to the shape, and rather smaller than the vignette. Place the whole in the light until the white margin is slightly tinted, or "blushed," as it sometimes called. The edges of the black mask should be slightly turned up or kept moving to prevent the junction of the tinting and the vignette being visible.

The above method represents a vignetted head on smooth grey paper, and is useful to show up the high lights on the face; but there is a modification of this effect, in which the appearance of a sketch on rough drawing-paper is produced.

Vignettes on Rough Drawing-paper.—If, instead of placing a piece of plain glass over the masked print, a thin negative of some diaper or pattern had been used, the design could have been printed on the paper instead of the even tint. A very good negative for this purpose is made as follows:—Obtain a sheet of the roughest drawing-paper, take a camel-hair brush dipped in thin sepia, and brush it evenly over the paper; the colour will fall into the depressions of the paper, and make the roughness still more visible. This should now be placed where a side light falls upon it, and photographed. A very thin negative is all that is required. This negative should be used in place of the plain glass, and, if not printed too dark, the effect of the delicate vignette inside the rough tint is very pleasing. It is better when using negatives for this purpose to place them in pressure-frames, instead of merely placing them or the print on the velvet board, to print, or perfect contact may not be obtained.

Medallions.—Medallions of oval and other forms are now a good deal used for small portraits. These are simply produced by gumming a mask, made of black or yellow paper, with an oval or other-shaped aperture, on to the negative, the mask preserving the part it covers white. These masks can be bought from the dealers cheaper and better than they can be made. Eccentric shapes are, usually, in bad taste; the oval and dome are quite sufficient for all purposes. If, instead of leaving the outside of the print—that protected by the mask—white, it could be tinted, the lights in the picture would have greater value, and the effect be improved.. To do this, the printed part should be covered with a black-paper disc corresponding with the mask used in printing, the print covered with glass, and exposed to the light until printed the required depth. In performing this operation it will be found convenient to gum the disc to the covering-glass. If texture could be added to this tinted margin, then another element of beauty would be added. This may be done in a similar manner to that described for vignettes, by using a negative made from rough drawing-paper; but, in this case, there is opportunity for a greater choice of objects from which to make the tinting negative, such as grained leather, marble of various kinds, paper-hangings—when suitable patterns can be obtained—and from the borders of old prints. In this, as in many other things connected with photography, there is a good deal of room for bad taste, which the photographer must try to avoid. He must remember that all these surrounding designs should assist the portrait, and not distract the attention from it.

Vignettes in Ornamental Borders.—The writer has lately produced some effects that have given much pleasure by using designs specially drawn for the purpose. The designs principally consist of an oval in the centre for the portrait, and a tablet underneath, on which the original of the portrait may sign his name. These forms are surrounded by flowers and other objects conventionally treated. The spaces for the portrait and name should be stopped out with black varnish, so as to print white. The easiest way to use these ornamental border negatives is as follows :—First print the border negative ; you will then have a print with a white oval space in the centre. Place this print on the portrait negative, taking care that it occupies the proper position in the oval. This is easily ascertained by holding the print and negative up to the light. It should then be placed in the frame and printed, care being taken that the vignetted gradation does not spread beyond its limits over the border.

There is a good deal of variety to be got out of the combination of the mask and vignette. Here is one of them.

Combination of Medallion and Vignette.—Vignette a head into the centre of the paper ; when this is done, place over it a black paper oval disc, taking care that the head comes in the centre under the mask. Place a piece of glass over the whole, and print. When the disc is removed, the print will represent a vignette surrounded by a dark oval. Many variations may be made of this form of picture, and there is much scope for skill and taste.

Any of the tinting negatives above described may be used, or they can be made from designs drawn on paper as we have already stated, or from natural objects. But if our reader has followed us clearly thus far, he is now in a position to form combinations for himself. This we recommend him to do, for there is an additional beauty in anything in art that indicates a distinctive style or shows thought and originality. There is too much tendency in portraitists to run in grooves, which the universal prevalence of the two styles, card and cabinet, help to promote. But we must caution the young photographer against the mistake of making changes for the sake of change. The "loud," and the bizarre, may attract foolish people, but it is only the beautiful that **will** secure the attention of the cultivated and refined.

CHAPTER XIV.

COMBINATION PRINTING.

THE scope of photography is wider than those who have only taken a simple portrait or landscape suppose. It is almost impossible to design a group that could not have been reproduced from life by the means our art places at our disposal. We do not mean to assert that such subjects as Michael Angelo's Last Judgment, or Raphael's Transfiguration, for instance, have ever been done in photography; but it is not so much the fault of the art, as of the artists, that very elaborate pictures have not been successfully attempted. It has not been the failing of the materials, unplastic as they are when compared with paint and pencils; it has been the absence of the requisite amount of skill in the photographer in the use of them, that will account for the dearth of great works in photography. The means by which these pictures could have been accomplished is Combination Printing, a method which enables the photographer to represent objects in different planes in proper focus, to keep the true atmospheric and linear relation of varying distances, and by which a picture can be divided into separate portions for execution, the parts to be afterwards printed together on one paper, thus enabling the operator to devote all his attention to a single figure or sub-group at a time, so that if any part be imperfect, from any cause, it can be substituted by another without the loss of the whole picture, as would be the case if taken at one operation. By thus devoting the attention to individual parts, independently of the others, much greater perfection can be obtained in details, such as the arrangement of draperies, the refinement of pose, and expression.

The most simple form of combination printing, and the one most easy of accomplishment and most in use by photographers, is that by which a natural sky is added to a landscape. It is well-known to all photographers that it is almost impossible to obtain a good and suitable sky to a landscape under ordinary circumstances. Natural skies are occasionally seen in stereoscopic slides and very small views ; but I am now writing of pictures, and not of toys. It rarely happens that a sky quite suitable to the landscape occurs in the right place at the time it is taken, and, if it did, the exposure necessary for the view would be sufficient to quite obliterate the sky ; and if this difficulty were obviated by any of the sun-shades, cloud-stops, or other inefficient dodges occasionally proposed, the movement of the clouds during the few seconds necessary for the landscape would quite alter the forms and light and shade, making what should be the sky— often sharp and crisp in effect—a mere smudge, without character or form. All these difficulties are got over by combination printing, the only objections being that a little more care and trouble are required, and some thought and knowledge demanded. The latter should be considered an advantage, for photographs, of a kind, are already too easy to produce. Of course, when a landscape is taken with a blank sky, and that blank is filled up with clouds from another negative, the result will depend, to a very great degree, upon the art knowledge of the photographer in selecting a suitable sky, as well as upon his skill in overcoming the mechanical difficulties of the printing. It is not necessary here to enter into a description of the art aspect of the matter, as that has often been discussed ; so we will confine ourselves to the mechanical details.

The landscape negative must have a dense sky, or, if it be weak, or have any defects, it must be stopped out with black varnish. In this case, it is better to apply the varnish to the back of the glass; by this means a softer edge is produced in printing than if painted on the varnished surface. With some subjects, such as those that have a tolerably level horizon, it is sufficient to cover over part of the sky while printing, leaving that part near the horizon gradated from the horizon into white.

It may here be remarked that in applying black varnish to the back of a negative, occasions will often be found where a softened or vignetted edge is required for joining, where a vignette glass or cotton wool cannot be applied; in such cases the edge of the varnish may be softened off by dabbing slightly, before it is set, with the finger, or, if a broader and more delicately gradated edge be required, a dabber made with wash-leather may be employed with great effect.

When an impression is taken, the place where the sky ought to be will, of course, be plain white paper; a negative of clouds is then placed in the printing-frame, and the landscape is laid down on it, so arranged that the sky will print on to the white paper in its proper place; the frame is then exposed to the light, and the landscape part of the picture is covered up with a mask edged with cotton wool. The sky is vignetted into the landscape, and it will be found that the slight lapping over of the vignetted edge of the sky negative will not be noticed in the finished print. There is another way of vignetting the sky into the landscape, which is, perhaps, better and more convenient. Instead of the mask edged with cotton wool, which requires moving occasionally, a curved piece of zinc or cardboard is used. Here is a section of the arrangement. The straight line represents the sky negative, and the part where it joins the landscape is partly covered with the curved shade. Skies so treated must not, of course, be printed in sunlight.

It is sometimes necessary to take a panoramic view. This is usually done, when the pantascopic camera is not employed, by mounting two prints together, so that the objects in the landscape shall coincide; but this is an awkward method of doing what could be much better accomplished by combination printing. The joining of the two prints is always disagreeably visible, and quite spoils the effect. To print the two halves of a landscape, taken on two plates, together, the following precautions must be observed: both negatives must be taken before the camera-stand is moved, the camera, which must be quite horizontal, pointing to one half of the scene for the first negative, and then turned to the remaining half of the view for the second negative. The two negatives should be obtained under exactly the same conditions of light, or they will not match; they should also be so taken that a margin of an inch and a-half or two inches is allowed to overlap each other; that is to say, about two inches of each negative must contain the same or centre portion of the scene. It is advisable, also, that they should be of the same density; but this is not of very great consequence, because any slight discrepancy in this respect can be allowed for in printing. In printing vignettes with cotton wool, or a straight-edged vignette glass, the edge of the left-hand negative on the side that is to join the other, taking care to cover up the part of the paper that will be required for the companion negative; when sufficiently printed, take the print out of the frame, and substitute the right-hand negative; lay down the print so that it exactly falls on the corresponding parts of the first part printed (this will be found less difficult, after a little practice, than it appears), and expose to the light, vignetting the edge of this negative, also, so that the vignetted part exactly falls on the softened edge of the impression already done. If great care be taken to print both plates exactly alike in depth, it will be impossible to discover the join in the finished print. If thought necessary, a sky may be added, as before described, or it may be gradated in the light, allowing the horizon to be lighter than the upper part of the sky.

Perhaps the greatest use to which combination printing is now put is in the production of portraits with natural landscape backgrounds. Many beautiful pictures, chiefly cabinets and card, have been done in this way by several photographers. The easiest kind of figure for a first attempt would be a three-quarter length of a lady, because you would then get rid of the foreground, and have to confine your attention to the upper part of the figure and the distance. Pictures of this kind have a very pleasing effect. In the figure negative, everything should be stopped out, with the exception of the figure, with black varnish; this should be done on the back of the glass when practicable, which produces a softer join; but for delicate parts—such as down the face—where the joins must be very close, and do not admit of anything approaching to vignetting, the varnish must be applied on the front. A much better effect than painting out the background of the figure negative is obtained by taking the figure with a white or very light screen behind it; this plan allows sufficient light to pass through the background to give an agreeable atmospheric tint to the distant landscape; and stopping out should only be resorted to when the background is too dark, or when stains or blemishes occur, that would injure the effect. An impression must now be taken which is not to be toned or fixed. Cut out the figure, and lay it, face downwards, on the landscape negative in the position you wish it to occupy in the finished print. It may be fixed in its position by gumming the corners near the lower edge of the plate. It is now ready for printing. It is usually found most convenient to print the figure negative first. When this has been done, the print must be laid down on the landscape negative so that the figure exactly covers the place prepared for it by the cut-out mask. When printed, the picture should be carefully examined, to see if the joins may be improved or made less visible. It will be found that, in many places, the effect can be improved and the junctions made more perfect, especially when a light comes against a dark—such as a distant landscape against the dark part of a dress—by tearing away the edge of the mask covering the dark, and supplying its place by touches of black varnish at the back of the negative; this, in printing, will cause the line to be less defined, and the edges to soften into each other. If the background of the figure negative has been painted out, the sky will be represented by white paper; and as white paper skies are neither natural nor pleasing, it will be advisable to sun it down.

If a full-length figure be desired, it will be necessary to photograph the ground with the figure, as it is almost impossible to make the shadow of a figure match the ground on which it stands in any other way. This may be done either out of doors or in the studio. The figure taken out of doors would, perhaps, to the critical eye, have the most natural effect, but this cannot always be done, neither can it be, in many respects, done so well. The light is more unmanageable out of doors, and the difficulty arising from the effect of wind on the dress is very serious. A slip of natural foreground is easily made up in the studio; the error to be avoided is the making too much of it. The simpler a foreground is in this case, the better will be the effect.

The composition of a group should next engage the student's attention. In making a photograph of a large group, as many figures as possible should be obtained in each negative, and the position of the joins so contrived that they shall come in places where they shall be least noticed, if seen at all. It will be found convenient to make a sketch in pencil or charcoal of the composition before the photograph is commenced. The technical working out of a large group is the same as for a single figure; it is, therefore, not necessary to repeat the details; but we give a reduced copy, as a frontispiece to this volume, of a large combination picture, entitled " When the Day's Work is Done,' ' by Mr. **H. P.** Robinson, a description of the progress and planning of which may be of use to the student.

A small rough sketch was first made of the idea, irrespective of any considerations of the possibility of its being carried out. Other small sketches were then made, modifying the subject to suit the figures available as models, and the accessories accessible without very much going out of the way to find them. From these rough sketches a more elaborate sketch of the composition, pretty much as it stands, and of the same size, 32 by 22 inches, was made, the arrangement being divided so that the different portions may come on 23 by 18 plates, and that the junctions may come in unimportant plates, easy to join, but not easy to be detected afterwards. The separate negatives were then taken. The picture is divided as follows :—

The first negatives taken were the two of which the background is composed. The division runs down the centre, where the light wall is relieved by the dark beyond it. These two negatives were not printed separately—it is advisable to have as few printings as possible—but were carefully cut down with a diamond, and mounted on a piece of glass rather larger than the whole picture, the edges being placed in contact, making, in fact, *one* large negative of the interior of the cottage, into which it would be comparatively easy to put almost anything. The next negative was the old man. This included the table, chair, and matting on which his feet rest. This matting is roughly vignetted into the adjoining ground of the cottage negative. The great difficulty at first with this figure was the impossibility of joining the light head to the dark background ; no amount of careful registration seemed equal to effect this difficult operation ; but if it could not be done, it could be evaded. Several clever people have been able to point out the join round the head, down the forehead, and along the nose, but we have never been able to see it ourselves, because we know it is not there. This is how the difficulty was got over. The figure was taken with a background that would print as nearly as possible as dark as the dark of the cottage. The join is nowhere near the head. but runs up the square back of the old woman's chair, then up the wall, and across the picture, over the head in an irregular line, and descends on the old man's back, whence it was easy to carry it down the dark edge of his dress and the chairs till it comes to the group of baskets, pails, &c., that fill up the corner. On the other side, the join runs along the edge of the table, and finds its way out where the floor coverings come together. The old lady was then photographed, and is simply joined round the edge; so also was the group in the corner, and the glimpse of the village seen through the window.

At first sight, it will appear difficult to place the partly-printed pictures in the proper place on the corresponding negative. There are many ways of doing this, either of which may be chosen to suit the subject. Sometimes a needle may be run through some part of the print, the point being allowed to rest on the corresponding part of the second negative. The print will then fall in its place at that point. Some other point has then to be found at a distance from the first; this may be done by turning up the paper to any known mark on the negative, and allowing the print to fall upon it; if the two separate points fall on the right places, all the others must be correct. Another way of joining the prints from the separate negatives is by placing a candle or lamp under the glass of the printing-frame —practically, to use a glass table—and throwing a light through the negative and paper; the join can then be seen through. But the best method is to make register marks on the negatives. This is done in the following manner. We will suppose that we wish to print a figure with a landscape background from two negatives, the foreground having been taken with the figure. At the two bottom corners of the figure negative make two marks with black varnish, thus ⌊⌋ ; these, of course, will print white in the picture. A proof is now taken, and the outline of the figure cut out accurately. Where the foreground and background join, the paper may be torn across, and the edges afterwards vignetted with black varnish on the back of the negatives. This mark is now fitted in its place on the landscape negative. Another print is now taken of the figure negative, and the white corner marks cut away very accurately with a pair of scissors. The print is now carefully applied to the landscape negative, so that the mark entirely covers those parts of the print already finished. The landscape is then printed in. Before, however, it is removed from the printing-frame, if, on partial examination, the joins appear to be perfect, two lead pencil or black varnish marks are made on the mark round the cut-out corners at the bottom of the print. After the first successful proof there is no need for any measurement or fitting to get the two parts of the picture to join perfectly ; all that is necessary is merely to cut out the little white marks, and fit the corners to the corresponding marks on the mask ; and there is no need to look if the joins coincide at other places, because, if two points are right, it follows that all must be so. This method can be applied in a variety of ways to suit different circumstances.

It is always well to have as few paintings as possible, and it frequently happens that two or more negatives can be printed together. For instance, the picture we have been discussing—"When the Day's Work is Done"—is produced from six negatives, but it only took three printings. The two negatives of which the cottage is composed was, as already explained, set up on a large sheet of glass, and printed at once ; the old man was also set upon another glass of the same size, with the negative of the glimpse through the window ; and the old woman was printed in like manner, with the corner group of baskets, &c. So that here were practically three negatives only. These were registered with corner marks so accurately that not a single copy has been lost through bad joins.

There are one or two things to consider briefly before concluding this subject.

It is true that combination printing—allowing, as it does, much greater liberty to the photographer, and much greater facilities for representing the truth of nature—also admits, from these very facts, of a wide latitude for abuse; but the photographer must accept the conditions at his own peril. If he find that he is not sufficiently advanced in knowledge of art, and has not sufficient reverence for nature, to allow him to make use of these liberties, let him put on his fetters again, and confine himself to one plate. It is certain (and this we put in italics, to impress it more strongly on the memory) that *a photograph produced by combination printing must be deeply studied in every particular, so that no departure from the truth of nature shall be discovered by the closest scrutiny.* No two things must occur in one picture that cannot happen in nature at the same time. If a sky is added to a landscape, the light must fall on the clouds and on the earth from the same source and in the same direction. This is a matter that should not be done by judgment alone, but by judgment guided by observation of nature. Effects are often seen, especially in cloud-land, very puzzling to the calm reasoner when he sees them in a picture; but these are the effects that are often best worth preserving, and which should never be neglected, because it may possibly happen that somebody will not understand it, and, therefore, say it is false, and, arguing still further on the wrong track, will say that combination printing always produces falsehoods, and must be condemned. A short anecdote may, perhaps, be allowed here. Some time ago a photograph of a landscape and sky was sent to a gentleman whose general judgment in art was admitted to be excellent; but he knew that combination printing was sometimes employed. In acknowledging the receipt he said, " Thank you for the photograph; it is a most extraordinary effect; sensational, certainly, but very beautiful; but it shows, by what it is, what photography cannot do; your sky does not match your landscape; it must have been taken at a different time of day, at another period of the year. A photograph is nothing if not true." Now it so happened that the landscape and sky were taken at the same time, the only difference being that the sky had a shorter exposure than the landscape, which was absolutely necessary to get the clouds at all, and does not affect the result. Another instance arose in connection with a picture representing a group of figures with a landscape background. Four of the figures were taken on one plate, at one operation; yet a would-be critic wrote at some length to prove that these figures did not agree one with another; that the light fell on them from different quarters; that the perspective of each had different points of sight; and that each figure was taken from a different point of view! These two cases are mentioned to show that it is sometimes a knowledge of the means employed, rather than a knowledge of nature—a foregone conclusion that the thing must be wrong, rather than a conviction, from observation, that it is not right—that influences the judgment of those who are not strong enough to say, "This thing is right," or "This thing is wrong, no matter by what means it may have been produced."

CHAPTER XV.

TONING THE PRINT.

If a print on albumenized paper be fixed without any intermediate process, the result is that the image is of a red, disagreeable tone, and unsightly. Moreover, it will be found that, if such a print be exposed to the atmosphere, it rapidly loses its freshness, and fades. In order to avoid this unsightliness, resort is had to toning, the toning, in reality, being the substitution of some less attackable metal for the metallic silver which forms a portion of the print. The usual metal used for substitution is gold applied in the state of the ter-chloride. It is not very easy to tell precisely how the substitution is effected; the question is, at present, sub judice, and, therefore, we propose to omit any theory that may have been broached. It is sufficient to say that it is believed the first step towards the reduction of the gold is the production of a hydrated oxide, and never metallic gold. Be that as it may, if a finely-divided silver be placed in a solution of chloride of gold, the silver becomes converted into the chloride, and the gold is quickly reduced to the metallic state ; and since gold combines with more chloride than does silver, it is manifest that when the substitution takes place,* the

---

* Silver subchloride and gold trichloride give silver chloride and gold.

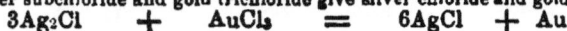

$$3Ag_2Cl \quad + \quad AuCl_3 \quad = \quad 6AgCl \quad + \quad Au$$

metallic gold deposited must be very much less than the silver. The colouring power of gold is, however, very great, when in the fine state of division in which we have it, being an intense purple to blue colour, and a very little of this mixed visually with the ruddy or brown colour of the albuminate which has been discoloured by light gives, after fixing, a pleasing tone. A picture, when toned thus, is composed of silver subchloride, metallic gold, and an organic compound of silver. If a print be kept in the toning bath too long, we are all aware that the image becomes blue and feeble, and the same disaster happens when a toning bath is too strong, i.e., is too rich in gold solution. The reason of this is, that too much gold is substituted for the silver in the sub-chloride, and there is in consequence too great a colour of the finely-precipitated gold seen. To make a toning bath, the first thing is to look after the gold. There is a good deal of chloride of gold sold, which is, in reality, not chloride of gold, but a double chloride of gold and of some such other base as potassium. and if it be paid for as pure chloride of gold, it is manifest that the price will be excessive. It is best to purchase pure chloride of gold, though it may be slightly acid, since subsequent operations correct the acidity. In our own practice we get fifteen-grain tubes, and break them open, and add to each grain one drachm of water, and in this state it is convenient to measure out. Thus, for every grain of gold to be used, it is only necessary to measure out one drachm into a measure. In delicate chemical operations, this would rightly be considered a rough method; but for a practical photographer it is sufficiently precise.

Now if chloride of gold alone were used, it would be found that the prints, after immersion in a dilute solution, were poor and "measley," and practice has told us that we must add something to the solution to enable it to act gradually and evenly. First of all, the gold solution must be perfectly neutral, and we know no better plan than adding to it a little powdered chalk, which at once neutralizes any free acid. It is not a matter of indifference what further retarder is added, for the reason that the more you retard the action, the more ruby-coloured becomes the gold, and less blue. A well-known experiment is to dissolve a little phosphorus in ether, and add it to a gallon of water, and then to drop in and stir about half a grain of chloride of gold. Phosphorus reduces the gold into the metallic state, but when so dilute the reduction takes places very slowly. The gold will, however, precipitate gradually, but it will be in such a fine state of division that it is a bright ruby colour. A very common addition to make to a toning bath is acetate of soda, and if the gold be in defect, the same appearance will take place in the solution. If chloride of lime, however, be added instead, and a commencement of precipitation of gold be brought about, the gold will be of a blue colour, having a slight tendency to purple. In this case, the grains of gold deposited are larger than when it is in the ruby state. The tone of the print then depends in a large measure on the degree of rapidity with which the gold is deposited. The quicker the deposit, the larger and bluer the gold, whilst an extremely slow deposition will give the red form. It often happens that no matter how long a print is immersed in a toning bath, it never takes a blue tone. The reason will be obvious from the above remarks.

We now give some toning baths which are much used.

No. 1.—Gold tri-chloride ...   .   1 grain
         Sodium carbonate ...   . 10 grains
         Water   ...   . 10 ounces

This bath must be used immediately after mixing, since the gold is precipitated by the carbonate. The tones given by this bath are purple and black. The prints should be toned to dark brown for the purple tone, and a slightly blue tone for the black tone.

No. 2.—Gold tri-chloride   ...   .   2 grains
        Saturated solution of chloride of lime   2 drops
        Chalk   ...   .   a pinch
        Water   ...   .   16 ounces

The saturated solution of chloride of lime is made by taking the common desinfecting powder, and shaking a teaspoonful up in a pint bottle. When the solids have settled, the clear liquid can be decanted off, and corked up till required. This is the solution used above. It is as well to keep this solution in the dark room.

The water with this bath should be hot (boiling better still), and the bath may be used when it is thoroughly cool. It is better, however, to keep it a day before using, since, when fresh, the action is apt to be too violent, and the prints are readily over-toned. The tone with this bath is a deep sepia to black. To get the first tone a very short immersion is necessary ; the prints should be almost red. For a black tone the prints should be left in the solution till they are induced to be of a purple hue.

No. 3 is made as follows :—

        Sodium acetate   ...   .   1 drachm
        Gold trichloride   ...   .   5 minim
        Distilled water   ...   .   12 ounces

This bath is a most excellent one in many respects, and should not be used under a week to get the best result. As this is a long time to keep a bath, it as as well to have two always on stock. It keeps indefinitely if proper care be taken of it. This produces a purple or brown tone, according to the length of time the print is immersed in it.

Now, as to toning the print. After the day's printing is done, the prints should be placed in a pan of good fresh water, in order to dissolve out all or a certain amount of silver nitrate that is invariably left in them. A puncheon, such as is used in dairies, is very convenient. It should be filled with water, and the prints placed in one by one, taking care that no one sticks to its neighbour, as this would be a fruitful source of unequal toning. Most water contains a little carbonate of lime and chloride of sodium, &c.; the water will therefore become milky. When the prints have been in the first water for ten minutes, they should be removed to another vessel of water, one by one. The first wash water should be placed in a wooden tub, with a tap let into it about six inches above the base, together with a little

Fig. 27.

common salt. The salt forms chloride of silver, which gradually precipitates, and the clear water is then drawn off on the next day, and the sediment is left undisturbed.

It now remains to see which toning bath is to be used. If No. 1 or 3, the whole of the free silver should as far as possible be washed away, which may entail three or four changes of water; the last two washings it will hardly repay to place in the tub; the second washing should certainly be added to it. If No. 2 toning bath be used, a little free silver should remain in the print; in fact, the washing should be confined to two changes of water.

When toning operations are commenced, the toning solution is poured off from any sediment that may be in the bottle containing it into a dish a couple of inches wider each way than the

largest print which has to be toned. If big prints have to be
toned, it is inadvisable to place more than a couple in the dish at
the same time, since there is a certain awkwardness in judging
of the amount of tone given to a print which is (say) between two
or three. The prints should be placed face up in the solution, and
great care should be taken that liquid separates each print from
the next one to it, otherwise there will be patches of unequal
toning. The dish containing the prints in the solution should
be gently rocked to secure a proper mixture of the solution
which may have been robbed of its gold in those strata next
surface of the prints. The rocking is also advisable to cause any
adhesion between two deep-toning prints impossible. If the prints
be of small size, a dozen or more may be toned at one operation.
Each print should be frequently brought to the surface of the
liquid, and examined in order to see how the toning action is
progressing. When one print is judged sufficiently toned, it is
removed to a dish containing pure water, and another untoned
print placed in the dish in its stead. This operation is continued
till all the prints are toned. We have heard that it has been
suggested to place the prints in water containing a littler acetic
acid or common salt, in order to stop the toning action con-
tinuing from the solution which may be held in the paper. The
former is most undesirable, acetic acid, as we shall see presently,
decomposing the fixing bath.

As to the addition of common salt, we can scarcely give a
favourable opinion regarding it. The addition of a chloride does,
in truth, alter the colour of the deposited gold (see ante), and it
may be this that gives rise to the opinion that it corrects toning
action. Of one thing we have little doubt, however, and that is,
that the addition of any large amount of common salt will tend
to turn the albumenate of silver into chloride, which in fixing
will materially weaken the print. When giving the formula of
the toning baths, we have indicated the depth to which toning
should take place. One great point to attend to is, that a print

should not be a slatey colour when fixed, and that can only be avoided by stopping the toning action when the print arrives at a blue-purple stage.

The toning bath, when used, should be replaced in the bottle, and we recommend that it be kept in a dark place, otherwise any chloride of silver which finds its way into the solution will darken and be a nucleus for the precipitation of gold from the solution. The energy of the toning bath would, in consequence, be wholly gone. It will be found that in very cold solutions formed in winter the toning action is much slower than in summer, and we need scarcely point out that this due to the fact that cold invariably retards chemical action. This retardation is not advantageous, and it will be found positively hurtful as to the colour of the precipitated gold. We therefore recommend that the toning solution and the dish in which it is to be poured should be warmed before the fire, the former to a temperature of about 70°F., and the latter a little higher. By this means the toning action will take place as rapidly as in warm weather, and the same tones be produced. It must be rememberd we are writing for all; not for those alone who have an elaborate arrangement for keeping their operating rooms at a good temperature in all weathers, but also for those who cannot afford the luxury. It is for this reason that we have given the above directions.

CHAPTER XVI.

SIR J. HERSCHEL was the first to point out that hyposulphite of soda would dissolve chloride of silver, and subsequently it has been found that it dissolves almost every organic salt of silver. In our early chapters we gave some examples of this. When we add hyposulphite to a salt of silver, such as the chloride, we get one of two reactions, the formation of a nearly insoluble double hyposulphite of soda and silver, or a readily soluble one.

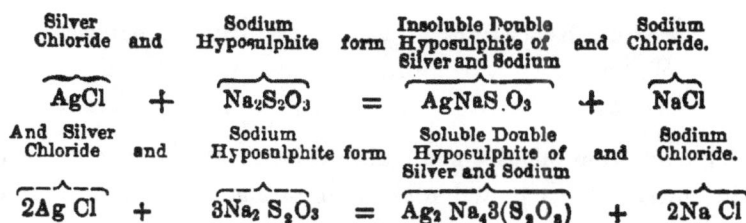

Silver Chloride	and	Sodium Hyposulphite	form	Insoluble Double Hyposulphite of Silver and Sodium	and	Sodium Chloride.
$AgCl$	$+$	$Na_2S_2O_3$	$=$	$AgNaS.O_3$	$+$	$NaCl$

And Silver Chloride	and	Sodium Hyposulphite	form	Soluble Double Hyposulphite of Silver and Sodium	and	Sodium Chloride.
$2AgCl$	$+$	$3Na_2S_2O_3$	$=$	$Ag_2Na_4 3(S_2O_3)$	$+$	$2NaCl$

The first insoluble double hyposulphite is formed when there is only a small quantity of sodium hyposulphite present; the soluble kind when the sodium hypsulphite is in excess. Since it is the soluble kind which we wish to form, it is manifest that the presence of a sufficiency of hyposulphite in the fixing bath is necessary. If not, we have left the insoluble form on the paper. If either of these two kinds of hyposulphite be made in a test-tube, we can readily simulate the effect of atmospheric exposure. If slightly acid water be added to the hyposulphite, it will be seen, when chloride of silver has been dissolved by the hyposulphite, that the precipitate or solution commences to blacken, sulphide of silver being formed. On the other hand, if we take albumenate of silver, and dissolve it in hyposulphite of soda, we shall find that the addition of acid gradually causes a yellow-looking compound to separate out, and it is probably this body formed in the paper which causes the gradual yellowing of the whites of silver prints.

What is taught us, then, by this observation is, that by thorough washing we must try and eliminate all traces of hyposulphite of silver, and, indeed, of the hyposulphite of soda, since the latter decomposes as rapidly in the presence of acid as does the silver compound.

The formula for the fixing bath which we recommend is :—

Sodium hyposulphite ...	...	...	...	4 ounces
Water ...	...	...	...	1 pint
Ammonia	...	...	...	½ drachm

The addition of the ammonia prevents any possibility of an acid reaction arising, and otherwise softens the film of albumen and the size of the paper, causing more rapid fixation and more thorough washing. Another thing the ammonia does is, that it prevents, in a great measure, blistering of the film of albumen, which is common in some highly-albumenized paper.

Experience has shown that one ounce of solid hyposulphite will fix with safety three sheets of paper, so that an idea can be formed of how much must be used for a day's printing. The hyposulphite bath which has been used one day should never be used the next, since it invariably contains the germs of decomposition in it from some cause or another. Indeed, the appearance of the solution indicates this is so, since it is usually of a yellow or brownish appearance.

The time required for fixing a print varies with the thickness of the paper used. As a rule, prints on the medium-sized paper require ten minutes' soaking in the bath, whilst thick-size requires fifteen minutes. Whilst toning, the dish containing the hyposulphite should be kept in a gentle rocking motion, as in toning, and for the same reasons. Prints may be examined from time to time, to see how the fixing progresses. When a print is not quite fixed, small spots of dark appearance will be seen when it is examined by transmitted light. The operation of fixing should be continued after these disappear for at least three or four minutes, in order that the hyposulphite of soda in the dish may get impregnated with the double silver and sodium salt which is in the print, and thus render washing more effectual. It should be noted that the dish for fixing should be at least as long and wide as the dish used for toning; that it should be deeper when, as a rule, all the prints are fixed at one time. Care should be taken that dishes which are used for sensitizing, toning, or fixing, *should not be used for anything else.* The glaze of porcelain dishes is often soft, and frequently absorbs a certain amount of the solutions used. Thus, if a porcelain dish be used for a solution of any aniline dye, it will often be found that it is permanently stained. Colour in this last is merely indication of what happens with any other solution. It will thus be seen that it is a mistake to use a dish or fixing when the glaze is cracked, since old hyposulphite must find its way into the body of the fresh solution that may be used, and thus institute a spontaneous decomposition, and a consequent want of permanence in the print. For our own part, we believe that a gutta-percha dish is a safer dish to use than any other, since it is impervious to any solution, and can be well scoured after fixing, and before being again brought into use. We believe that much of the fading of prints may be traced to the use of unsuitable dishes for fixing.

CHAPTER XVII.

WASHING THE PRINT.

THERE are very many apparatus designed for washing prints; but we believe that, where few prints have to be treated, careful hand-washing is as superior to machine-washing, as hand-made paper is to machine-made. In our own practice we take the prints from the fixing-dish, and immerse them in a large puncheon of water, and allow them to soak for five minutes, after which we carefully pour off all the water, and replenish with fresh, in which we leave them for a quarter-of-an-hour. After that we take the prints and place them on a glass slab, and, with a squegee, squeeze as much water as possible out of each separately; this we repeat twice. After two more washings of a quarter-of-an-hour, we then wash for half-an-hour, and, with a sponge, dab them as dry as possible, and again immerse for half-an-hour. After repeating this operation twice, we allow a stream of running water to pour into the puncheon for a couple of hours, carrying the stream through an india-rubber pipe, at the end of which is a glass tube, to the bottom of the puncheon, and so that the pour of water goes against the side. By this means there is a constant stir in the water, and the water flows over the edge of the puncheon. It is convenient to cut a notch in the top rim of the puncheon, so that the water may find an exit before reaching the level of the rim. The prints are then taken out, sponged once more, and dried. By this arrangement we have got prints which are perfectly unfaded, though they have been in existence eighteen years, and have been to the tropics, and in the dampest climates. This method of washing, though tedious, should be applied to all prints; but, in the present day, it can hardly be hoped that it can be immediately adopted, on account of the attention it requires; we therefore describe an apparatus which can be used. It was designed by Mr. England, and consists of a working trough, as shown in the figure, which is automatically worked by an overshot wheel. We need not enter into the details of the invention, as they are self-evident.

Fig. 28.

This washing arrangement causes the prints to be alternately soaking in water, and draining. Whilst in the water they are perpetually being shaken apart by the movement of the tray, and thus every part of the print gets washed, and it is almost impossible for two prints to stick together. In all washing apparatus there is a danger of air-bells forming on the surface of the prints while in the water, but in this form there is the advan-

tage that whilst draining the air-bells must break, and so water
on rising to the level of the prints can obliterate any of the evil
effects which would be caused by their being perpetually remaining on one spot. It is useless to attempt to describe other forms
of the apparatus, since there are so many; we have chosen one
which appears to us to be a satisfactory form.

The following tests for the elimination of hyposulphite are
taken from another work of this series.*

"The following is a most delicate test.

"Make the following test solution:—

Potassium permanganate	... ...	2 grains
Potassium carbonate	... ...	20 ,,
Water		1 quart

"The addition of a few drops of this rose-coloured solution to
a pint of water will yield a slightly pink tinge. If there be any
trace of sodium hyposulphite present, this colour will give place
to one of a greenish hue.

"If the permanganate be not at hand, the following well-
known starch iodide test may be adopted:—

"Take about two drachms of water and a small piece of starch
about the size of a small pea; powder and boil the starch in the
water till the solution is quite clear; add one drop of a saturated
solution of iodine in alcohol to this clear liquid. It will now
become dark blue. Of this solution drop two drops into two
clean test tubes, and fill up one with distilled water and the
other with the water to be tested; a faint blue colour should
be perceptible in the first test tube. In the second test tube,
should hyposulphite be present, this blue colour will have
disappeared, the iodide of starch becoming colourless in its
presence. The best mode of comparing the two waters is by
placing a piece of white paper behind the test tubes.

---

* "Instruction in Photography," 4th edition.

H

"It frequently occurs that though sodium hyposulphite can-not be detected in the washing water, it may be present in the paper itself. The paper on which most prints are taken being sized with starch, if a *very* weak solution of iodine be applied with a brush across the *back* of a print, a blue mark will indicate the *absence* of the hyposulphite. Care must be taken that the iodine solution is *very* weak, otherwise a part of the iodine will first destroy the trace of the salt, and then the remainder will bring out the blue re-action."

We finish this chapter by quoting our maxims to be observed in printing.

" *Maxims for Printing.*

"1. The prints should have the highest lights *nearly* white, and the shadows verging on a bronzed colour before toning.

"2. Place the prints, before toning, in the water, face down-wards, and do not wash away too much of the free nitrate of silver.

"3. The toning solution must be neutral or slightly alkaline, and not colder than 60°.

"4. Tone the prints to purple or sepia, according as warm or brown prints are required.

"5. Move the prints, in both the toning and fix ng solutions, repeatedly, taking care that no air-bubbles form on the surface.

"6. Take care that the fixing bath is not acid.

"7. Use fresh sodium hyposulphite solution for each batch of prints to be fixed.

"8. Wash thoroughly after and before fixing.

"9. Make a sensitizing bath of a strength likely to give the best results with the negatives to be printed.

"10. Print in the shade, or direct sunshine, according to the density of the negative."

CHAPTER XVIII.

PRINTING ON PLAIN PAPER.

RINTS on plain paper are sometimes of use; for instance, they form an excellent basis on which to colour. They are of course duller than an albumenized print, since the image is formed more in the body of the paper than on the surface. The following formula may be used :—

Ammonium chloride... ...	60 to 80 grains
Sodium citrate	100 ,,
Sodium chloride	20 to 30 ,,
Gelatine	10 ,,
Distilled water	10 ounces

Or,

Ammonium chloride	100 grains
Gelatine	10 ,,
Water	10 ounces

The gelatine is first swelled in cold water, and then dissolved in hot water, and the remaining components of the formulæ are added. It is then filtered, and the paper is floated for three minutes, following the directions given on page 10. If it be required to obtain a print on plain paper in a hurry, a wash of citric acid and water (one grain to the ounce) may be brushed over the back of ordinary albumenized paper, and, when dried, that side of the paper may be sensitized and printed in the ordinary manner. For cold tones the wash of the citric acid may be omitted.

The toning and fixing are the same as described in Chapters XII. and XIII.

CHAPTER XIX.

PRINTING ON RESINIZED PAPER.

THE following is taken from another volume of this series.*

To Mr. Henry Cooper we are indebted for a valuable printing process, founded on substituting resins for albumen or other sizing matter. The prints obtained by this process are very beautiful, and lack that gloss of albumen which is often called vulgar and inartistic.

The following are the two formulæ which Mr. Cooper has communicated to the writer :—

Frankincense...	•••	•••	•••	10 grains
Mastic... ...	•••	•••	...	8 ,,
Calcium chloride	...	•••	5 to 10	,,
Alcohol ...	...	•••	...	1 ounce

When the resins are dissolved in the alcohol, the paper is immersed in the solution, then dried and rolled. The sensitizing bath recommended is as follows (though the strong bath given at page 126 will answer) :—

Silver nitrate...	•••	•••	...	60 grains
Water ...	•••	•••	...	1 ounce

---

* "Instruction in Photography," 4th edition.

To the water is added as much gelatine as it will bear without gelatinizing at 60° Fah.

The second formula gives very beautiful prints, soft and delicate in gradation. . .

The paper is first coated with an emulsion of white lac in gelatine, which is prepared as follows :—

3 ounces of *fresh* white lac are dissolved in 1 pint of strong alcohol, and after filtering or decanting, as much water is added as it will bear without precipitating the lac; 1 ounce of good gelatine is soaked and dissolved in the pint of boiling water, and the lac solution is added with frequent stirring. If, at any stage of this operation, the gelatine is precipitated, a little more hot water must be added. The pint of lac solution ought, however, to be emulsified in the gelatine solution.

To use the emulsion, it is warmed, and the paper immersed in or floated on it for three minutes. When dry, the coated surface is floated in the following for a couple of minutes :—

Ammonium chloride ...	...	...	10 grains
*Magnesium lactate ...	...	...	10 ,,

When dry, it is sensitized on a moderately strong bath (that given on the last page will answer).

If more vigour in the resulting prints be required, it is floated on—

Citric acid ...	...	...	...	5 grains
White sugar ...	...	...	...	5 ,,

This last bath improves by use, probably by the accumulation of silver nitrate from the sensitized paper.

---

* Or ten minims of ammonium lactate.

Any of the toning baths given in Chapter XII. will answer, though Mr. Cooper recommends:—

Solution of gold tri-chloride (1 gr. to 1 dr. of water)   2 dr.
Pure precipitated chalk   ...   .   a pinch
Hot water   ...   . 10 ounces

2 dr. of sodium acetate are to be placed in the stock-bottle, and the above solution filtered on to it.   This is made up to 20 ounces, and is fit for use in a few hours ; but it improves by keeping.

In commencing to tone, place a few ounces of water in the dish, and add an equal quantity of the stock solution, and if the toning begins to flag a little, add more of it from time to time.

With the resin processes over-toning is to be carefully avoided.

Resinized paper may be obtained from most photographic dealers, we believe, and for some purposes is an admirable substitute for albumenized paper.

CHAPTER XX.

PRINTING ON GELATINO-CHLORIDE EMULSION PAPER.

MR. W. T. WILKINSON has recently brought forward the notion of using gelatine instead of albumen as a medium for holding the silver chloride in printing.   He uses the following formula :—

Barium chloride	...	...	...	2,440 grains
Gelatine	...	...	...	2,000 ,,
Water	...	...	...	20 ounces

The gelatine is allowed to swell in the water, and, by the aid of heat, is dissolved; the barium chloride is then added.   Next he prepares—

Silver nitrate	...	...	...	1,700 grains
Water	...	...	...	5 ounces

and adds this to the former, little by little, in a large bottle with much shaking, or pours it slowly into the former in a large jar, stirring briskly the whole time.   This makes an emulsion of silver chloride, and is used without washing.   When required for use, the gelatine, which will have set when cold, is swelled by placing the jar containing it in hot water, and is then transferred to a dish.   The dish should be kept warm by being placed, supported on small blocks, in a tin tray (about two inches larger in dimensions every way than the dish) filled with hot water, the temperature of which should be about 150° F. to commence with. Saxe or Rive paper may be coated by rolling the sheet face outwards, and placing the edge of the roll upon the gelatine. The two corners of the paper in contact with the solution are then taken hold of by the fingers, and raised. The paper will unroll of itself, and take up a thin layer of the gelatine emulsion. The sheet of paper is then suspended to dry. All these operations are, of course, conducted in the dark room. The behaviour of the paper in the printing-frame is precisely the same as albumenized paper, and the washing and toning are conducted in the same way. For a fixing bath is used—

Sodium hyposulphite ...	...	...	2 ounces
Water ..	...	...	...	... 20 ,,

The washing after fixing is more rapid than with albumenized paper. It is washed in ten or twelve changes of water for ten minutes, and then placed for five minutes in an alum bath made as follows :—

Potash alum ...	...	...	...	5 ounces
Water ...	...	...	...	... 20 ,,

The print is washed in a few changes of water, and the prints are ready for drying and mounting. The advantage of the alum bath is that the hyposulphite is destroyed into harmless products, and the gelatine is rendered insoluble by it. In the formula given there is large excess of chloride, and we recommend that instead of using 2,440 grains of barium chloride, 2,050 grains be used. (Mr. Wilkinson has used that amount of the barium salt that would be required exactly to convert 1,700 grains of silver nitrate into silver chloride, if the formula for barium chloride were $BaCl_3$ instead of $BaCl_2$.) It will be seen that whichever formula is used, there is no silver left to combine with the gelatine, and hence the image will be entirely formed by metallic silver, and not an organic salt of silver.

CHAPTER XXI.

DRYING THE PRINTS.

In many establishments the prints are taken direct from the washing water, and hung up by American clips, and thus allowed to dry. When this is done, the prints curl up as the water leaves the paper, and they become somewhat unmanageable. If prints have to be dried at all before mounting—and they must, unless they are trimmed before toning—a better plan is to make a neat heap of some fifty or sixty of the same size (say cartes), place them on blotting-paper, and drain for a time, and then in a screw-press (such as is used to press table-cloths, for instance) to squeeze out all superfluous water. After a good hard squeeze the prints should be separated, and the plan adopted by Mr. England carried out. He has frames of light laths made, of about 6 feet by 3 feet, and over this frame is stretched ordinary paperhanger's canvas. The prints are laid on this to dry spontaneously, and they cockle up but very little. The frames, being light, are easily handled. After the squeezing is done, supposing the room in which they are placed be not very damp or very cold, the prints will be ready for trimming and mounting in a couple of hours. To our minds there is nothing superior to this mode of drying, since the squeezing in the press tends to eliminate every slight trace of hyposulphite which might be left in them.

Trimming the Prints.—Perhaps more prints are ruined in trimming than in any other way, when the operator is inexperienced, since it requires judgment to know which part of the print to trim off, so that a right balance shall be kept. In trimming landscape prints, it is impossible to give any set rules; the judgment as to what is artistic must be the guide. Of one thing we may be certain, that, unless the operator who took the original negative knows exactly how to balance his picture on the focussing-screen, the print will always bear cutting down in one direction or the other. Such a clipping, of course, alters the size of the print, which, if it be one of a series, will be a misfortune; but, on the other hand, the artistic value of the individual print will be increased.

For portraits there are some few rules which should be followed in trimming. Always allow the centre of the face to be a little "out" from the central line of the print, making more space on the side towards which the sitter is looking. Allow a carte or cabinet to be cut in such a way that, if the sitter has been leaning on something, it does not seem as if he had been leaning on nothing. Should there be an unintentional lean on the part of the sitter, trim the print so that he appears in an upright position.

To trim the print, there should be the various sized shapes in glass used. Thus there should be glasses with bevelled edges for the carte, the cabinet, and other sizes, which can be laid on the print as a guide to the trimming. The absolute trimming may be done either by shears or by a knife, a leather cutters' knife being excellent, since it is rounded, and can be brought to a keen edge very readily. When the knife is used, the print is placed on a large glass sheet of good thickness, the pattern placed over it, and, whilst this is held down by the left hand, the knife is used by the right, keeping it close to the edge of the pattern glass. When shears are used, the print is held against the pattern glass by the left hand, and each side trimmed by one clip, taking care to make the cut parallel to the edges of the pattern glass. It requires a little practice to prevent clipping the glass as well as the paper, but for small sized prints, such as the carte, the shears have a decided advantage over the knife.

For cutting out ovals, Robinson's trimmer is an excellent adjunct to the mounting-room, and in this case ovals stamped out of sheet brass are used as guides.

The figure will show the action of the trimmer. The small

Fig. 29.

wheel is the cutter, and, being pivotted, it follows the curve against which it is held. It is better to cut out prints with this trimmer on sheet zinc in preference to glass, the edge of the wheel being kept sharp for a longer time than where the harder glass is used. To use the trimmer, the print is placed on the sheet of zinc, the oval mask (or square mask, with slightly rounded corners) is placed in position on it. The wheel of the trimmer is brought parallel to, and against, the edge of the mask, the handle being grasped by the right hand, the thumb to the left, and the fingers on the right. A fairly heavy downward pressure is brought to bear on the trimmer, and at the same time the wheel is caused to run along the edge of the mask. The cut should be clean, and the join perfect, if proper care be taken. It is desirable to practise on ordinary writing paper before it is taken into use for prints. Square masks with very slightly rounded corners can be used ; the smaller the wheel, the less curved the corners need be. It will be seen that there is a limit to smallness of the wheel used, since, if too small, the stirrup on which it is pivoted would rest upon the mask. The larger the wheel the easier is the cutting.

With larger sizes than the carte or the cabinet, mounting may often have to be delayed, since it is easier to keep a stock of unmounted prints (say landscapes) unmounted than it is when they are mounted. In this case the prints should be put away as flat as possible. The plan of drying we have indicated takes out the "curl," but even then they will not be flat enough to be handily put away. We therefore recommend the practice of stroking the prints. A flat piece of hard wood, about 1 foot long and 1½ inch broad, and the thickness of a marquoise scale, has its edges carefully rounded off. The print is seized by one corner in one hand and unrolled; the face of the print is brought in contact with a piece of plate glass. The "stroker," held by the other hand, is brought with its rounded edge on to the back of the print near the corner held by the first hand. Considerable pressure is brought upon the stroker, and the print is drawn through between it and the plate. The print is then seized by another corner and similarly treated. By this means a gloss is put upon the print, and the creases and cockles are obliterated. The print is now ready for trimming.

It is well to have a square of glass with true edges cut to the size of the pictures. The prints should be trimmed upon a sheet of plate glass, a sharp penknife being used to cut them. A rough test for ascertaining if the opposite sides are equal is to bring them together, and see if both corners coincide.

It may sometimes be found useful to cut out a print into an oval. The following method for tracing any ellipse may be employed :—On a thickish piece of clean paper draw a line A B, making it the *extreme* width of the oval required. Bisect it at O, and draw D O C at right angles to A B. Make O C equal to *half* the smallest diameter of the ellipse. With the centre C and the distance O B, draw an arc of a circle, cutting A B in E and F. Place the paper on a flat board, and at E and F fix two drawing-pins. Take a piece of thread and knot it together in such a manner that half its length is equal to A F. Place the thread round the two pins at E and F, and stretch it out to tightness by the point of a lead pencil. Move the pencil guided by the cotton, taking care to keep it upright. The resulting

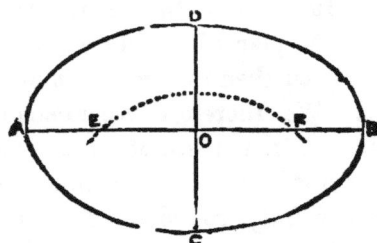

Fig. 30.

figure will be an ellipse. Modifications of this figure may be made by making a second knot beyond the first knot, and placing the point of the pencil in the loop formed. When the figure has been traced in pencil on paper, it should be carefully cut out with a sharp penknife, and placed on the print which is to be trimmed into an oval. When so placed, a faint pencil line is run round on the print, and the cutting out proceeds either by scissors or penknife.

CHAPTER XXII.

MOUNTING PHOTOGRAPHS.

THERE are many photographers who, unfortunately, are quite indifferent as to the medium they use in mounting the trimmed photographs. So long as the medium will cause the adherence of the back of the print to the cardboard employed, they are perfectly satisfied, whether it be paste fresh or sour, or starch or gelatine in a similar condition. If any of our readers have had the misfortune to have their rooms papered with rancid paste, they will have noticed that the unpleasant smell attending it has not been removed from the room for weeks, and that there is a liability of the return of the disgusting odour when the air is at all damp. In this case the fact that decomposition is going on is detected by the olfactory nerves, because the quantity is considerable. It is none the less true, however, that every square inch of the surface of the wall paper is undergoing the same ordeal, and that if it contains any colour, &c., which would be affected by decomposing organic matter, there would be but small chance of the paper retaining its fresh appearance. Were a silver print mounted with the same paste, we need scarcely point out that danger to its permanency is to be apprehended. Paste, we know, is as a rule tabooed, but there is no occasion for it to be so if care be taken that it is absolutely fresh when employed in mounting. In looking for a mounting material, we should

endeavour to find something which does not readily take up moisture. Glue, gelatine, dextrine, and gum are all inadmissible on this account; on the other hand, starch, arrowroot, cornflour, and gum tragacanth, when once dry, do not seem to attract moisture.

Referring to glue, Mr. W. Brooks says* that he has recently seen many photographs which have been mounted with that medium, and in some cases, where the glue has been put on too thickly, it swells up into ridges, showing marks of the brush with which it is applied, and each ridge after a time turns brown. The same writer is not wholly in favour of starch, but in our own opinion pure white starch is as good a material as can be met with. To prepare it for use as a mountant, a large teaspoonful of starch is placed in the bottom of a cup, with just sufficient cold water to cover it. This is allowed to remain for a couple of minutes, after which the cup is filled with boiling water, and well stirred; the starch should then be fairly thick, but not so thick as to prevent a brush taking up a proper supply for a good sized print. We will suppose that we are going to mount a day's work of carte-de-visite prints. In a former chapter we have said that it is desirable that the prints should be left damp. If they are dried, they should be *slightly* moistened, and placed in a heap one above the other, as by so doing the moisture is confined, and one damping of all the prints is sufficient. In our own practice we have, as is natural, all the prints with the faces downwards. A stiff bristle brush is then dipped into the pot containing the starch, and the starch brushed over the back of the top print. This one is then carefully raised from the print beneath it, and, supposing it to have been properly trimmed, it is laid upon the card, and pressed down by means of a soft cloth, and placed on one side to dry. The next print is then treated in the same manner, and so on. By this plan no starch gets on

---

* See Mr. W. Brooks' article in Photographic Almanac, 1881.

the face of the prints, which is a desideratum. With a little practice, just sufficient starch will be brushed on each carte, and no more. Young hands, however, are sometimes apt to give more than a fair share to them ; in this case, after pressing the print down with the soft cloth, it may be useful to place on the print a piece of writing paper, and press all superfluous starch out by a rounded straight-edge, or an ivory or wooden paper knife. The card in this case should be placed on a slab of thick glass, so as give an even pressure. The starch, which will exude beyond the edges of the card, should be carefully wiped off with a *clean* cloth.

This is of course a method to be adopted only in the case of bungling mounting, but it is useful then, and may save a carte. It should be remembered that the less mounting medium used, the greater is the chance of a silver print not fading.

To mount larger prints, the back should be slightly damped, and the brush with the starch applied with cross strokes, so that every part is covered. Particular care should be taken that the corners and edges are not missed, since it often necessitates re-mounting the print, which is to be avoided as far as possible, since it is a troublesome matter. The rounded-edged ruler, and the sheet of white paper, is also useful here, since over a large surface there is more difficulty in getting even layers of starch, than over smaller ones. When a print has to be mounted with a margin, the places where the top corners have to come should be marked with a fine pencil point. By a little dexterity, the top edge of the print, the back of which has been covered with starch in the manner described, can be brought into the position indicated by these dots, and be then lowered without puckers or folds. It should be remembered that the print should just cover the pencil marks, since it is almost impossible to erase blacklead with india-rubber, if any starch should by accident get on it.

It is well to dry these prints under pressure, since the cardboard is apt to cockle. A couple of boards rather longer than the prints suffice for the purpose. The mounted prints are laid between them, a sheet of clean blotting-paper separating each, and a few weights placed on the top board. For prints of moderate size, a table-cloth press is an excellent substitute.

As to the kind of mounts to be used, opinions vary. To our mind, the simpler they are, the better they look. It is not rare to find a regular advertisement of the photographic establishment below a carte or cabinet print. To say the least of it, this is bad taste, and we are sure it is bad art. If the work be good, it needs no recommendation; and if it be bad, the less of an advertisement that appears, the better it is for the photographer. At the back of a carte or cabinet is the place where any advertisement should appear; but even here it may be overdone. When we find the back of the carte got up with any amount of gold-lettering flourishes, and no blank space on which the eye can rest without encountering some one especial merit of the artist, we may expect to find on the front of the card the same kind of tawdry work. It is seldom advisable to have the mount of a white colour, though for cartes or cabinets, in which the margin will be hidden in the album, this is not of much consequence; but for prints in which the margin shows, it is generally advisable to have some slight tint visible, preferably of a cream or buff colour. There are some classes of work which will, however, bear a white margin, but it is rarely the case; and we advise, as a general rule, that there should be some tone on it, to prevent its attracting the eye away from the picture by its whiteness. Black mounts are much in vogue at the present time, and they are effective and artistic; but chemical analysis has shown them not to be safe, since they are enamelled with substances which are apt to induce fading. A good and stable black mount is a desideratum, which it is to be hoped will be found before long.

Notwithstanding our preference for starch as a mountant, we

I

give a method of preparing glue for the same purpose. The glue used should be light, and as clean as possible. It should be shredded and soaked in sufficient clean water to cover it for five or six hours; any dust which may have adhered to it will find its way into the water. The water should be poured off and replaced by an equal quantity of fresh. The vessel containing it is heated over a small gas jet or spirit lamp until solution takes place. The liquid is then thinned down with warm water till it is of proper consistency, a point which is soon learned by a little practice. An ordinary small glue pot will be found convenient.

It is sometimes useful to have at hand a mounting solution which will not cockle the mount, and the late Mr. G. Wharton Simpson gave a formula which is very good in this respect. Fine cut gelatine or shredded glue is swollen in the least possible quantity of water, and this is boiled with alcohol, with much stirring. If 80 grains of Nelson's No. 1 fine cut gelatine are taken, 3 dr. of water should be used for making it, and to it 2 oz. of alcohol be added. When cool this sets into a jelly, and can be used by letting the bottle into which it has been transferred stand in hot water. Prints can be mounted on foolscap paper with this medium without any serious cockling being apparent.

It should be recollected that no two batches of paper will mount exactly alike, some expanding more than others. It is well to mount a trial print before doing many, to see exactly how the paper under manipulation behaves.

Rolling the Prints.—After the prints have been under the hands of the retoucher, they should be rolled in a rolling-press in order to give a brightness to the printed image. It would be invidious to point out any particular press that should he used. Suffice it to say, there are many excellent ones in the market. The directions for cleaning and using the press are supplied with each machine; we therefore refrain from saying anything about them.

CHAPTER XXIII.

DEFECTS IN PRINTS.

THE bath solution is sometimes repelled by the paper, and this is found chiefly in highly albumenized paper, and is generally caused by the paper being too dry. Passing the sheet of paper over the steam from a saucepan will generally effect a cure.

Small white spots, with a black central pin-point, are often met with in prints. Dust on the paper during sensitizing will cause them, the grit forming a nucleus for a minute bubble. All paper should be thoroughly dusted before being floated on the sensitizing bath.

Grey, star-like spots arise from small particles of inorganic matter, such as ferric oxide, lime, &c., h are present in the paper. They become more apparent by decomposition during the printing operations. They may generally be discernible by examining the paper by transmitted light.

Bronze lines (straight) occur through a stoppage during floating the paper in the sensitizing solution. Should the lines be irregular, forming angles and curves, it is probable that a scum of silver oxide, &c., may be detected on the surface of the sensitizing solution. A strip of blotting-paper drawn across the bath will remove the cause of the defect.

Should the print appear marbled, it may be surmised that the sensitizing solution is weak, or that the paper has not been floated sufficiently. In some cases it may arise from imperfect

albumenizing; but in ordinary commercial samples the cause
can be easily traced.

Red marks on the shadows may appear during toning, and are
very conspicuous after fixing. They generally arise from handling the paper with hot, moist fingers after sensitizing; greasy
matter being deposited on the surface, prevents the toning bath
acting properly on such parts.

Weak prints are generally caused by weak negatives. Such
can be partially remedied by paying attention to the strength of
the sensitizing bath (see Appendix), and by using washed
paper.

Harsh prints are due to harsh negatives. They can generally
be remedied by paying attention to the mode of printing, as
given in Chapter IX. If the negative be under-exposed and
wanting in detail, there is, however, no cure for this defect.

A red tone is due to insufficient toning; whilst a poor and
blue tone is due to an excess of toning.

The whites may appear yellow from imperfect washing,
imperfect toning, imperfect fixing, or from the use of old sensitized paper.

Should prints refuse to tone, either the gold has been ex-
hausted, or else a trace of sodium hyposulphite has been carried
into the toning bath by the fingers or other means. A trace of
hyposulphite is much more injurious to the print than a fair
quantity of it. Should the toning bath refuse to tone after the
addition of gold, it may be presumed that it is contaminated by
a trace of sodium hyposulphite.

A dark mottled appearance in the body of the paper indicates
imperfect fixing, combined with the action of light on the
unaltered chloride during fixing. If the fixing bath be acid, the
excess of acid combines with the sulphur, and forms hydrosulphuric acid, which will also cause the defect.

The cause of mealiness or "measles" in the print has been
explained in page 32.

CHAPTER XXIV.

ENCAUSTIC PASTE.

THE value of an encaustic paste in improving the effect of photographic prints has become very generally recognised amongst photographers. A good encaustic confers three special benefits on the print: it gives depth, richness, and transparency to the shadows; it renders apparent delicate detail in the lights which would otherwise remain imperceptible; and it aids in protecting the surface, and so tends to permanency. One of the writers has in his possession prints that were treated with an encaustic paste thirteen years ago, which retain all their original freshness and purity, while prints done at the same time from the same negatives have gone, to say the least of it, ".off colour."

Various formulæ for the preparation of encaustic pastes have been published, and many of them very excellent. The qualities required are, easiness of application, and the capacity of giving richness and depth without too much gloss, and of yielding a hard, firm, permanent surface. For a proper combination of all these qualities, nothing has ever approached the paste of the late Adam-Salomon, of which the following is the formula:—

Pure white wax	...	...	...	500 grains
Gum elemi	...	...	...	10 ,,
Benzole	...	...	...	200 ,,
Essence of lavender	...	...	...	300 ,,
Oil of spike	...	...	...	15 ,,

The wax is cut into shreds, and melted in a capsule over a water bath. Placing it in a jar, and the latter in a pan of hot water, will serve. Powder the elemi, and dissolve it in the solvent, using gentle heat. Some samples of elemi are soft and tough, and will not admit of powdering, in which case it may be roughly divided into small portions, and placed in a bottle with the solvents. Strain through muslin, and add the clear solution to the melted wax, and stir well. It is then poured into a wide-mouthed bottle, and allowed to cool.

The encaustic paste is put on the prints in patches, and then rubbed with a light, quick motion, with a piece of flannel, until a firm, fine surface is obtained.

We give another simple formula which is efficient, though we ourselves prefer the above.

> White wax cut into shreds ... ... 1 ounce
> Turpentine ... ... 1 ,,

and thinned down, if necessary, till it has the consistency of "cold cream."

Yet another is—

> White wax ... ... 1 ounce
> Benzole ... ... 2 ounces.

CHAPTER XXV.

ENAMELLING PRINTS.

THERE are several modes of enamelling prints, but there is none better than that described by Mr. W. England, which we quote in his words. " I have a glass having a good polished surface (patent plate is not necessary), and rub over it some powdered French chalk tied up in a muslin bag. Dust off the superfluous chalk with a camel's hair brush, and coat with enamel collodion. I find it an improvement to add to the collodion usually sold for the purpose 2 dr. of castor oil to the pint. When the collodion is well set, immerse the plate in a dish of water. When several prints are required to be enamelled, a sufficient number of plates may be prepared and put in dishes ; this will save time. Now take the first plate, and well wash under a tap till all greasiness has disappeared ; place it on a levelling stand, and pour on as much water as the plate will hold. Then lay the print on the top, squeeze out all the water, and place the plate and print between several thicknesses of blotting-paper to remove all superfluous moisture. The plate, with the print in contact, should now be placed in a warm room to dry spontaneously, when the print will come easily from the glass. Care should be taken not to attempt to remove the print till quite dry.

If the pictures required to be enamelled have been dried, it will be necessary to rub over them some ox-gall with a plug of soft rag; otherwise the water will run in globules on the surface, and make blisters when laid on the collodion.

"I may mention that prints done in this way lose their very glossy surface on being mounted, but retain their brilliancy, which I think is an improvement, as I dislike the polished surface usually given to the print when gelatine is employed."

CHAPTER XXVI.

CAMEO PRINTS.

At one time there was a rage amongst photographers to produce cameos. and, for this purpose, a special piece of apparatus was required to produce the embossing. The figure will explain it.

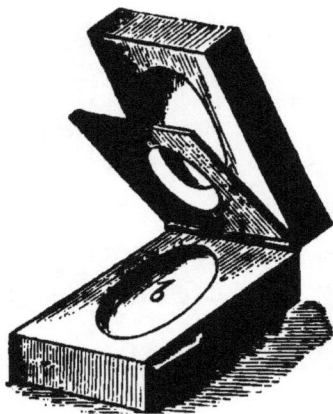

Fig. 31.

The print, after mounting, was enamelled by coating a plate with collodion—as described above—and a thin film of liquid gelatine applied. In some cases the carte itself was gelatinized, dried, and damped, and placed in contact with the collodion film. The carte was placed face downwards on the gelatine, and placed under pressure till quite dry. It was then removed, and bore on its surface a high gloss caused by the collodion. It was then ready for embossing, which was effected by placing it in the above apparatus.

Some people like the style; and it will be seen that great variety in it may be made by printing sufficient depth of border round the cameo; but, for our own part, we think that, in an art point of view, they are decidedly vulgar; and besides which, the surface of the cameo is readily scratched, since it is raised. We only give a brief account of what has been done in this direction, not to encourage its adoption, but rather to caution the photographer.

APPENDIX.

REDUCTION OF OVER-EXPOSED PRINTS.

MR. ENGLAND writes as follows to the Photographic Journal, and we can unhesitatingly say that the method of reducing an over-printed proof is excellent.

" A simple and certain method of reducing over-printed proofs has been one of the wants long felt by all photographers. It is well known that in every photographic establishment even the most careful printers cannot always be sure of getting the exact depth of tone required, and proofs occasionally get over-printed. Of course prevention is better than cure; but, when a remedy is necessary, the method I am about to describe answers admirably. I tried a great many experiments before I succeeded to my satisfaction. I found that cyanide of potassium totally destroyed the print, even when used moderately strong. By using a weaker solution it was well under control, and the exact depth could be readily obtained; but during the washing to remove the cyanide the action of the latter continued, and spoiled every proof. I then tried several methods to arrest the action of the cyanide, but without success. It then occurred to me to use the cyanide in such a weak state that but little should be held in the paper, only sufficient to reduce the print to the required depth; for this purpose I made a bath of only four drops of saturated solution of cyanide to a pint of water. The prints immersed at first showed

no signs of getting lighter, but after about an hour the most perfect results had been obtained with prints considerably over-
printed. With lighter pictures a less time is required. roofs
treated in this way lose nothing of their tone during the after-
washing, which should be thoroughly done, and, when dry, retain all the brilliancy of an ordinary print."

The plan of using cyanide has, we know, often been proposed,
but with no success until, we believe, Mr. W. Brooks gave a
formula which worked successfully with him.

Another plan, proposed by Mr. L. Warnerke, for effecting the
same thing is the use of ferric sulphate. A weak solution is
prepared, and the print immersed in it. The reduction takes
place rapidly, but evenly.

We need scarcely say that it is better not to have to use
either of these remedies, by avoiding over-printing; but as
mistakes will occur, it is evident that the above will be of use at
times.

UTILIZATION OF SILVER RESIDUES.

All paper or solutions in which there is silver should be saved,
as it has been proved by experience that from 50 to 75 per cent.
of the whole of the silver used can be recovered by rigid adherence to the careful storage of " wastes."

1. All prints should be trimmed, if practicable, before toning
and fixing; in all cases these clippings should be collected.
When a good basketful of them is collected, these, together with
the bits of blotting-paper attached to the bottom end o
sensitized paper during drying, and that used for the draining
of plates, should be burnt in a stove, and the ashes collected.
These ashes will naturally occupy but a small space in
comparison with the paper itself. Care should be taken that
the draught from the fire is not strong enough to carry up the
ashes.

2. All washings from prints, waters used in the preparation of

dry plates, all baths, developing solutions (after use), and old toning baths, should be placed in a tub, and common salt added. This will form silver chloride.

3. The old hyposulphite baths used in printing should be placed in another tub. To this the potassium sulphide of commerce may be added. Silver sulphide is thus formed.

4. To No. 1 nitric acid may be added, and the ashes boiled in it till no more silver is extracted by it. The solution of silver nitrate thus produced is filtered off through white muslin, and put aside for further treatment, when common salt is added to it to form chloride, and added to No. 2.

5. The ashes may still contain silver chloride. This may be dissolved out by adding a solution of sodium hyposulphite, and adding the filtrate No. 3.

6. No. 2, after thoroughly drying, may be reduced to metallic silver in a reducing crucible* by addition of two parts of sodium carbonate and a little borax to one of the silver chloride. These should be well mixed together, and placed in the covered crucible in a coke fire, and gradually heated. If the operator be in possession of one of Fletcher's gas furnaces he can employ it economically, and with far less trouble than using the fire. (It is supplied with an arrangement for holding crucibles, which is useful for the purpose.) After a time, on lifting off the cover, it will be found that the silver is reduced to a metallic state. After all seething has finished, the crucible should be heated to a white heat for a quarter of an hour. The molten silver should be turned out into an iron pan (previously rubbed over with plumbago to prevent the molten metal spirting), and immersed in a pail of water. The washing should be repeated till nothing but the pure silver remains.

The silver hyposulphite, having been reduced to the sulphide

---

* The crucible should be of Stourbridge clay.

by the addition of the potassium sulphide, is placed in a crucible, and subjected to a white heat; the sulphur is driven off, and the silver remains behind.

Another method of reducing silver chloride to the metallic state is by placing it in water slightly acidulated with sulphuric acid together with granulated zinc. The zinc is attacked, evolving hydrogen, which, in its turn, reduces the silver chloride to the metallic state, and forming hydrochloric acid. After well washing, the silver may be dissolved up in nitric acid.

Yet another method is to take sugar of milk and a solution of crude potash, when the silver is rapidly reduced. This requires careful washing, and it is well to heat the metal to a dull red heat to get rid of any adherent and insoluble organic matter which may have been formed, before dissolving it in nitric acid.

To Print from Weak and Hard Negatives.

Should a negative be found very hard, a slight modification of the sensitizing solution will be found beneficial, supposing the ordinary paper is to be used.

Silver nitrate ... ... 30 grains
Water ... ... 1 ounce

The negative should in this case be printed in the sun. The more intense the light, the less contrast there will be in the print, as the stronger light more rapidly effects a change in the albuminate than if subjected to weaker diffused light. The reason for the reduction in quantity of the silver nitrate in the solution is given on page 15.

To print from a weak negative, the sensitizing solution should be :—

Silver nitrate ... ... 80 grains
Water ... ... 1 ounce

The printing should take place in the shade; the weaker the negative, the more diffused the light should be.

If a negative be dense, but all the gradations of light and shade be perfect, the strong bath, and, if, possible, a strongly-salted paper, should be used. The printing should take place in sunlight.

To Make Gold Tri-Chloride [AU CL₃].

Place a half-sovereign (which may contain silver as well as copper) in a convenient vessel ; pour on it half a drachm of nitric acid, and mix with it two-and-a-half drachms of hydrochloric acid ; digest at a gentle heat, but do not boil, or probably the chlorine will be driven off. At the expiration of a few hours add a similar quantity of the acids. Probably this will be sufficient to dissolve all the gold. If not, add acid the third time ; all will have been dissolved by this addition, excepting, perhaps, a trace of silver, which will have been deposited by the excess of hydrochloric acid as silver chloride. If a precipitate should have been formed, filter it out, and wash the filter paper well with distilled water. Take a filtered solution of ferrous sulphate (eight parts water to one of iron) acidulated with a few drops of hydrochloric acid, and add the gold solution to it; the iron will cause the gold alone to deposit as metallic gold, leaving the copper in solution. By adding the gold solution to the iron the precipitate is not so fine as if added *vice versa*. Let the gold settle, and pour off the liquid; add water, and drain again, and so on till no acid is left, testing the washings by litmus paper. Take the metallic gold which has been precipitated, re-dissolve in the acids as before, evaporate to dryness on a water bath (that is, at a heat not exceeding 212° F.) The resulting substance is the gold tri-chloride. To be kept in crystals this should be placed in glass tubes hermetically sealed. For non-commercial purposes it is convenient to dissolve it in water (one drachm to a grain of gold). Ten grains of gold dissolved yield 15·4 grains of the salt. Hence if ten grains have been dissolved, 15·4 drachms of water must be added to give the above strength.

To Make Silver Nitrate.

Silver coins are mostly alloyed with tin or copper. In both cases the coin should be dissolved in nitric acid diluted with twice its bulk of water. If tin be present there will be an insoluble residue left of stannic oxide. The solution should be evaporated down to dryness, re-dissolved in water, filtered, and again evaporated to dryness. It will then be fit for making up a bath. If copper be present, the solution must be treated with silver oxide.

The silver oxide thus formed is added, little by little, till the blue or greenish colour has entirely disappeared. This will precipitate the copper oxide from the copper nitrate, setting free the nitric acid, which, in its turn, will combine with the silver oxide. The copper will fall as a black powder mixed with any excess of silver oxide there may be. Take one or two drops of the solution in a measure, and add a drachm of water, and then add ammonia to it till the precipitate first formed is re-dissolved. If no blue colour is apparent, the substitution of the silver for the copper is complete ; if not, more silver oxide must be added till the desired end is attained. Distilled water must next be added till the strength of the bath is that required. This can be tested by the argentometer.

If to a solution of silver nitrate a solution of potash be added, a precipitate will be formed. This is the silver oxide. The potash should be added till no further precipitation takes place. The oxide should be allowed to settle, the supernatant fluid be decanted off (a syphon arrangement is very convenient), and fresh distilled water added to it. This, in its turn, after the oxide has been well stirred, should be decanted off. The operation should be repeated five or six times, to ensure all nitrate of potash being absent, though its presence does not matter for a printing bath, since this or some other nitrate is formed when the paper is floated.

THE END.

www.ingramcontent.com/pod-product-compliance
Lightning Source LLC
Chambersburg PA
CBHW021934190326
41519CB00009B/1021